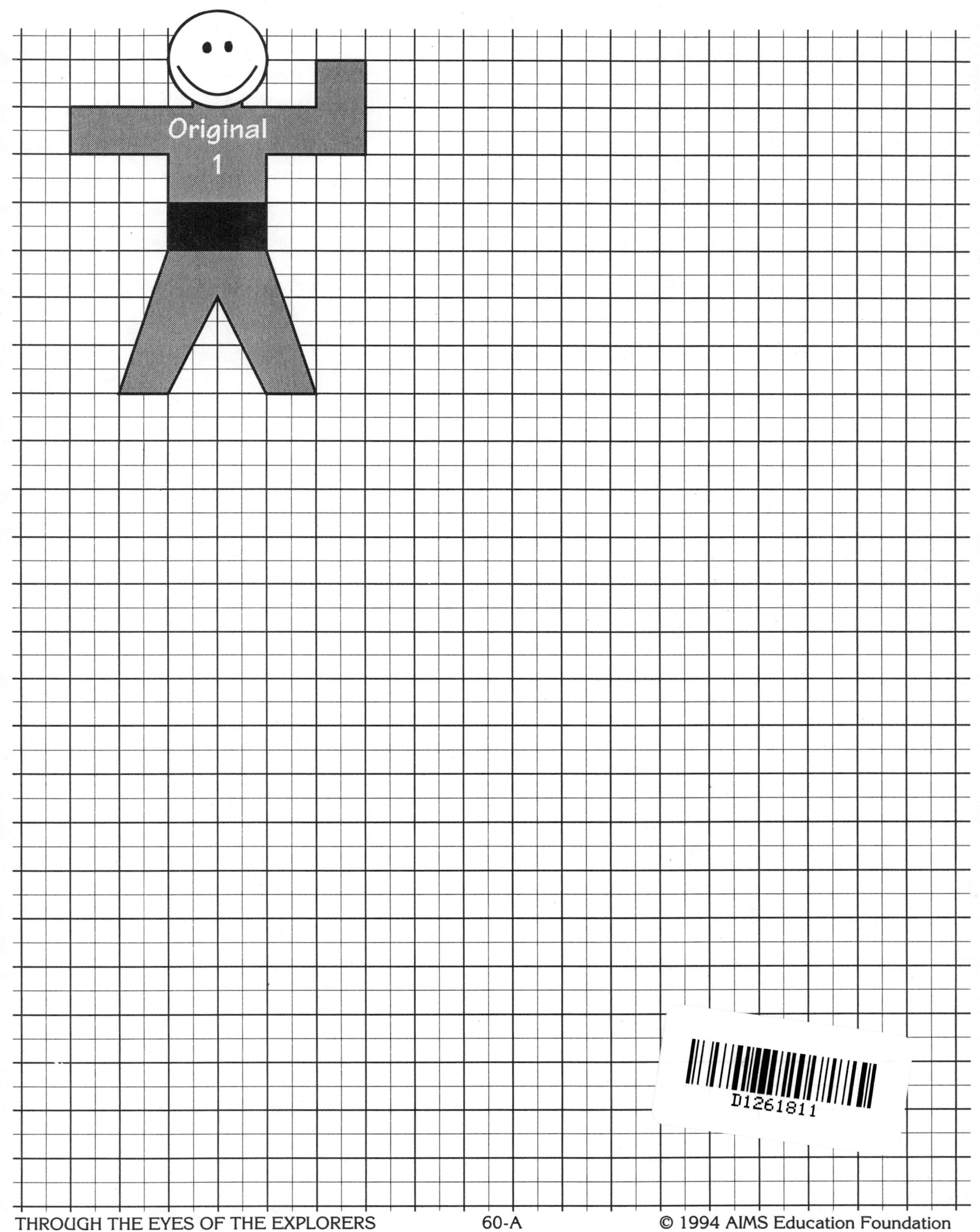
Original
1
D1261811

Minds-on Math and Mapping

Author

Sheldon Erickson

Editors

Betty Cordel

Judith Hillen

Illustrator

Brenda Dahl

Content Consultant

Dr. Fareed Nader
Professor of Surveying Engineering
California State University
Fresno, CA

This book contains materials developed by the AIMS Education Foundation. **AIMS** (**A**ctivities **I**ntegrating **M**athematics and **S**cience) began in 1981 with a grant from the National Science Foundation. The non-profit AIMS Education Foundation publishes hands-on instructional materials (books and the monthly *AIMS* magazine) that integrate curricular disciplines such as mathematics, science, language arts, and social studies. The Foundation sponsors a national program of professional development through which educators may gain both an understanding of the AIMS philosophy and expertise in teaching by integrated, hands-on methods.

ISBN **1-881431-48-7**

Printed in the United States of America

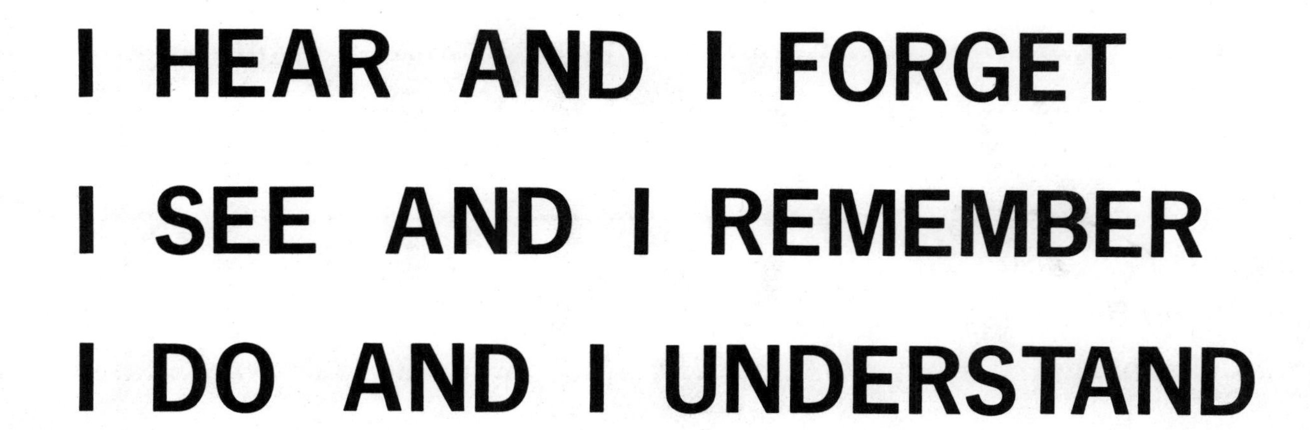
I HEAR AND I FORGET
I SEE AND I REMEMBER
I DO AND I UNDERSTAND

–Chinese Proverb

Table of Contents

THROUGH THE EYES OF THE EXPLORERS

Minds-on Math and Mapping

Introduction

Mapping has been a dynamic adventure of the mind, an attempt to understand our awesome world. It provides a mathematically rich situation in which the application of mathematical skills are used to communicate a visual form. Maps are never made for the demonstrations of mathematical power, but to illustrate dimensions about how far, how high, how big, how much.... Mapping deals with a broad range of mathematical topics including: proportional reasoning, measurement, and geometry. The internal connections made within the discipline naturally integrate these areas to provide students with a more coherent, holistic understanding of the various concepts being studied.

Maps reduce our multi-dimensional world to fit on small, flat surfaces. This reduction, however, gives rise to distortion. It has been the quest of cartographers to create maps with as little distortion as possible. If students are to contend with these distortions, we must provide them with opportunities to take part in the exciting adventure of constructing maps and actually experiencing these distortions in real-world settings.

In this book, students are given the opportunity to construct maps from first-hand information and primary sources. As a result, they will not only learn how to interpret maps, but will also become aware of their shortcomings. The investigations are divided into five units. The mathematical concepts of the units get progressively more complex allowing the students to enter at a level that is appropriate for them.

Four of the five units begin within an historical context which deals with the mapping of the United States for its westward expansion. The historical context may be introduced at the beginning of the unit to provide motivation. After completing the investigations in the unit, the historical component may be completed as an assessment of the students' understandings of the concepts within that unit.

The investigations allow for the development of interdisciplinary units. History, Science, and Math teachers can all find valuable experiences for their students while helping them to recognize the interwoven nature of their learning.

Through the Eyes of the Explorers: Minds-on Math and Mapping

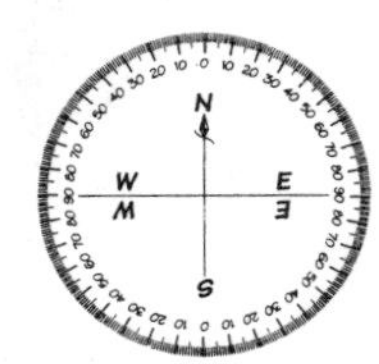

Direct Measure Maps

Diary Daze
It's a Court Case
"Honey, I shrunk the...."
Pace Race
Pacing the Yard

In this section, students measure areas around the school and apply proportional reasoning to reduce the measurements to make scale drawings. The measurements are taken with tape measures and pacing. Students apply their acquired skills to determine the trail of an early pioneer from his writings and measurements.

Maps from Global Position

Overland Mapping
Treasure Hunt
Backward Mapping
Sun Dance
Nearly North
Where on Earth?

By using a compass to make and follow treasure maps students learn about headings and angles. From observing the changes in the sun's position, students find out how to determine the coordinates of longitude and latitude for their position. Through these experience they come to appreciate the rigors of a surveyor as they use the sightings of Captain Fremont to construct a map of the Oregon Trail.

Indirect Measures

Sizing Up Shadows
Paring It Down to Size
Background Blockers
Tunnel Vision
End-Sights
Measuring Up
Going Ballistic

To make measurements by triangulation, students must be grounded in their understanding of the proportional relationships of similar triangles. These investigations progress from a direct comparison of corresponding parts to the development of the abstract concept of the trigonometric functions.

Indirect Measure Maps

A Mapping Expedition
Target Practice
Plane Table Mapping
Out-Landish Mapping

Students apply their understanding of indirect measurements to make maps.They learn to use transits and plane tables to construct maps as explorers did. Using the sightings and headings made by Lewis and Clark, students reproduce the map of the Columbia River made by Clark.

Topographical Maps

Space Maps
Topping Off Mount St. Helens
Uncanny Vision
On the Level
Hill Sides
Major Mountain Mapping

Students learn several ways to display three dimensional information on the two dimensional surface of a map. They make measurements of elevations from which they construct a topographical map of the area. Using patterns to decipher the digital data received from a space probe, students construct a map to show the dimensionally of the object.

Through the Eyes of Explorers Concept Map

Maps show the positions of geographic locations and the distances between them.

Linear measures are taken between geographic locations.

Linear measures may be taken by direct measurement.

Linear measures may be taken by indirect measurement.

Indirect measurement is based on establishing similar triangles.

The ratios of sets of corresponding sides of all similar triangles are equal.

Proportions are used to determine the unknown lengths of similar triangles.

Surveyors establish similar triangles by using angle-side-angle.

The distance between objects is reduced by some proportion to allow them to be drawn at a scale smaller than the actual size.

Position of a geographic location may be established by its angle of separation relative to another landmark and a distance.

Position of a geographic location may be established by its compass heading relative to a landmark, and the distance from the landmark.

Compass headings deviate depending on the position on the earth.

Maps may represent more than two dimensions. Other dimensions might include data on elevation, temperature fields, atmospheric pressure, or vegetation.

Topographical maps use contour lines to represent changes in elevation.

Contour lines mark points of equal elevation.

Elevation is recorded relative to a reference elevation such as sea level.

One way to measure elevations is by sighting angles between points.

Elevations may be taken directly by sounding, or indirectly by time elapsed for waves to travel (sound, light, radio).

Position of a geographic location may be established by determining the longitude and latitude.

Longitude and latitude can be determined by finding how far stars have moved relative to the Prime Meridian and the Equator in a time period.

Longitude and latitude are angles measured relative to the positions of the Prime Meridian and Equator on the globe.

INTEGRATED PROCESSES

	Observing	Comparing and Contrasting	Collecting and Recording Data	Interpreting Data	Predicting/ Inferring	Applying/Generalizing/ Concluding	Modeling
Diary Daze	X		X	X	X	X	
It's a Court Case	X		X	X		X	
"Honey, I Shrunk the..."	X		X	X			
Pace Race	X		X	X	X	X	
Pacing the Yard	X		X	X			X
Overland Mapping		X		X			
Treasure Hunt	X			X		X	
Backward Mapping	X		X	X			
Sun Dance	X	X	X		X	X	
Nearly North	X		X	X		X	
Where on Earth?	X		X			X	
Sizing Up Shadows	X		X	X		X	
Paring It Down to Size	X	X	X			X	
Background Blockers	X	X	X			X	
Tunnel Vision	X	X	X			X	
End-Sights	X	X	X			X	
Measuring Up	X		X	X		X	
Going Ballistic	X		X	X			
A Mapping Expedition	X	X		X		X	
Target Practice	X		X	X		X	
Plane Table Mapping	X		X	X		X	
Out-Landish Mapping	X		X			X	
Space Maps	X	X		X			X
Topping Off Mount St. Helens	X	X				X	
Uncanny Vision	X		X	X	X	X	
On the Level	X	X	X	X		X	
Hill-Sides	X		X	X		X	
Major Mountain Mapping	X		X	X		X	

MATH CONCEPTS

	Averaging	Estimating	Measuring	Proportional Reasoning	Scaling	Trigonometry	Using Charts and Graphs	Patterns/functions	Integers	Graphing	Slope	Geometry/ Geometric Constructions	Spatial Visualization	Topographical Mapping
Diary Daze			X		X									
It's a Court Case			X	X										
"Honey, I Shrunk the..."			X	X										
Pace Race		X	X					X		X	X			
Pacing the Yard			X	X	X									
Overland Mapping					X					X				
Treasure Hunt		X	X											
Backward Mapping			X		X									
Sun Dance								X						
Nearly North			X									X		
Where on Earth?			X				X		X					
Sizing Up Shadows	X		X	X				X		X				
Paring It Down to Size			X	X				X		X				
Background Blockers			X	X				X		X				
Tunnel Vision			X	X				X		X				
End-Sights			X	X				X		X				
Measuring Up			X	X		X		X		X				
Going Ballistic	X		X	X		X	X							
A Mapping Expedition			X	X	X									
Target Practice			X	X	X							X		
Plane Table Mapping			X	X								X		
Out-Landish Mapping			X	X	X							X		
Space Maps					X			X		X			X	X
Topping Off Mt. St. Helens					X								X	X
Uncanny Vision		X	X							X			X	X
On the Level			X		X				X				X	X
Hill-Sides			X	X	X				X	X	X	X	X	X
Major Mountain Mapping			X	X								X	X	X

Project 2061 Benchmarks

- *Important contributions to the advancement of science, mathematics, and technology have been made by different kinds of people, in different cultures, at different times.*
- *Mathematics is helpful in almost every kind of human endeavor–from laying bricks to prescribing medicine or drawing a face. In particular, mathematics has contributed to progress in science and technology for thousands of years and still continues to do so.*
- *Technology is essential to science for such purposes as access to outer space and other remote locations, sample collection and treatment, measurement, data collection and storage, computation, and communication of information.*
- *Because the earth turns daily on an axis that is tilted relative to the plane of the earth's yearly orbit around the sun, sunlight falls more intensely on different parts of the earth during the year. The difference in heating of the earth's surface produces the planet's seasons and weather patterns.*
- *Some changes in the earth's surface are abrupt (such as earthquakes and volcanic eruptions) while other changes happen very slowly (such as uplift and wearing down of mountains).*
- *The motion of an object is always judged with respect to some other object or point and so the idea of absolute motion or rest is misleading.*
- *Information can be carried by many media, including sound, light, and objects. In this century, the ability to code information as electric currents in wires, electromagnetic waves in space, and light in glass fibers has made communication millions of times faster than is possible by mail or sound.*
- *Most computers use digital codes containing only two symbols, 0 and 1, to perform all operations. Continuous signals (analog) must be transformed into digital codes before they can be processed by a computer.*
- *Numbers can be written in different forms, depending on how they are being used. How fractions or decimals based on measured quantities should be written depends on how precise the measurements are and how precise an answer is needed.*
- *The expression a/b can mean different things: a parts of size 1/b each, a divided by b, or a compared to b.*
- *Graphs can show a variety of possible relationships between two variables. As one variable increases uniformly, the other may do one of the following: always keep the same proportion to the first, increase or decrease steadily, increase or decrease faster and faster, get closer and closer to some limiting value, reach some intermediate maximum or minimum, alternately increase and decrease indefinitely, increase and decrease in steps, or do something different from any of these.*
- *Some shapes have special properties. Triangular shapes tend to make structures rigid, and round shapes give the least possible boundary for a given amount of interior area. Shapes can match exactly or have the same shape in different sizes.*
- *The graphic display of numbers may help to show patterns such as trends, varying rates of change, gaps, or clusters. Such patterns sometimes can be used to make predictions about the phenomena being graphed.*
- *It takes two numbers to locate a point on a map or any other flat surface. The numbers may be two perpendicular distances from a point, or an angle and a distance from a point.*

- *The scale chosen for a graph or drawing makes a big difference in how useful it is.*
- *Organize information in simple tables and graphs and identify relationships they reveal.*
- *Read simple tables and graphs produced by others and describe in words what they show.*
- *Mathematical ideas can be represented concretely, graphically, and symbolically.*
- *Scale drawings show shapes and compare locations of things very different in size.*
- *Estimate distances and travel times from maps and the actual size of objects from scale drawings.*
- *Mathematical statements can be used to describe how one quantity changes when another changes. Rates of change can be computed from magnitudes and vice versa.*
- *Read analog and digital meters on instruments used to make direct measurements of length, volume, weight, elapsed time, rates, and temperature, and choose approximate units for reporting various magnitudes.*

Benchmarks for Science Literacy
American Association for the Advancement of Science

NCTM Standards

- *Value the role of mathematics in our culture and society*
- *Use the skills of reading, listening, and viewing to interpret and evaluate mathematical ideas*
- *Develop common understandings of mathematical ideas, including the role of definitions*
- *See mathematics as an integrated whole*
- *Use mathematics as an integrated whole*
- *Reflect on and clarify their own thinking about mathematical ideas and situations*
- *Validate their own thinking*
- *Develop and apply a variety of strategies to solve problems, with emphasis on multistep and nonroutine problems*
- *Develop formulas and procedures for determining measures to solve problems*
- *Develop, analyze, and explain methods for solving proportions*
- *Use computation, estimation, and proportions to solve problems*
- *Generalize solutions and strategies to new problem situations*
- *Analyze functional relationships to explain how a change in one quantity results in a change in another*
- *Use patterns and functions to represent and solve problems*
- *Apply mathematical thinking and modeling to solve problems that arise, such as art, music, psychology, science, and business*
- *Use problem-solving approaches to investigate and understand mathematical content*
- *Make inferences and convincing arguments that are based on data analysis*
- *Explore problems and describe results using graphical, numerical, physical, algebraic, and verbal mathematical models or representations*
- *Verify and interpret results with respect to the original problem situation*
- *Systematically collect, organize, and describe data*
- *Understand and apply reasoning processes, with special attention to spatial reasoning and reasoning with proportions and graphs*
- *Describe and represent relationships with tables, graphs, and rules*
- *Model situations using oral, written, concrete, pictorial, graphical, and algebraic methods*
- *Estimate, make, and use measurements to describe and compare phenomena*
- *Extend their understanding of the concepts of perimeter, area, volume, angle measure, capacity, and weight and mass*
- *Understand the structure and use of systems of measurement*
- *Extend their understanding of the process of measurement*
- *Understand and apply ratios, proportions, and percents in a wide variety of situations*
- *Compute with whole numbers, fractions, decimals, integers, and rational numbers*
- *Understand, represent, and use numbers in a variety of equivalent forms (integer, fraction, decimal, percent, exponential, and scientific notation) in real-world and mathematical problem situations*
- *Investigate relationships among fractions, decimals, and percents*
- *Develop number sense for whole numbers, fractions, decimals, integers, and rational numbers*
- *Understand and appreciate the need for numbers beyond the whole numbers*
- *Develop an appreciation of geometry as a means of describing the physical world*
- *Understand and apply geometric properties and relationships*
- *Represent and solve problems using geometric models*

Curriculum and Evaluation Standards for School Mathematics
National Council of Teachers of Mathematics

Supply List

This list may help you in assembling the supplies for the activities you have chosen.

- adding machine tape
- bamboo skewers
- board-1"X12"X10" (minimum)
- butcher paper
- calculators
- carbon paper
- chalk
- circular bubble level
- clay
- clear packing tape
- clipboard
- colored markers
- corrugated cardboard
- cubes-2 cm
- dowels (1/4"), sharpened
- dowel-1/2"
- drawing compasses
- EMT Conduit-1/2"
- easel paper-1" grid
- empty coffee cans with plastic lids
- flashlights
- food coloring
- glue
- graph paper
- hammer
- large cans
- magnetic compass
- masking tape
- meter sticks
- nails
- overhead transparency film
- paint
- paper
- paper cups
- pennies
- potatoes
- protractors
- push pins
- rulers
- scissors
- straightedge
- straws
- string
- tagboard
- tape measure-1.5 meter
- tape measure-30 meter
- tape measure-100 ft
- tongue depressors
- toy action figures
- vinyl tubing-1/4"
- wood sticks

Diary Daze

Topic
Direct measure maps

Key Question
Using a map and John B. Wyeth's account of his journey, can you determine where he went on his travels?

Focus
Students will use the information from an historical journal to determine the position of its author.

Guiding Documents
NCTM Standards

- *Develop and apply a variety of strategies to solve problems, with emphasis on multistep and nonroutine problems*
- *Value the role of mathematics in our culture and society*
- *Estimate, make, and use measurements to describe and compare phenomena*

Project 2061 Benchmarks

- *Important contributions to the advancement of science, mathematics, and technology have been made by different kinds of people, in different cultures, at different times.*
- *Estimate distances and travel times from maps and the actual size of objects from scale drawings.*

Math
Logic
Measurement
  length
Scaling
Rates

Integrated Processes
Observing
Collecting and recording data
Interpreting data
Inferring
Generalizing

Materials
Maps
Ruler
String
Masking tape
Colored markers

Background Information

This 1832 account of a fur trappers' rendezvous by John Wyeth is of historical interest. It was the largest rendezvous of its kind and represented the peak of this period of western expansion.

Students doing this activity will gain insight to historical interpretation. By using the journal entry as a primary source of information, students will see the need to make interpretive decisions. For example, does the author mean July 1, 1832 when he writes "about the first of July." If he does mean the first day of the month, the rest of the chronology does not work. Does he mean June 30?...June 29?...28? In their roles as historians, the students will need to make decisions that make the most sense. This decision making provides a valuable experience for the students to see that history is an interpretive process. This activity also provides the students with insights into why early maps were not accurate when compared to our present knowledge.

The Wyeth expedition generally followed the route of the Oregon Trail until they crossed the Rocky Mountains in the vicinity of South Pass or "the highest part or ridge of the mountain." From this point, they left the future "emigrant road" and followed the Green River, crossing over to the Snake River, across the Tetons through Teton Pass, and on to Pierre's Hole which was located on the western slope of the Tetons near the present-day border of Idaho and Wyoming.

Names of landmarks have changed since this account was written. The La Platte has been shortened to Platte. The river referred to as the Colorado of the West is presently called the Green River. Trappers had rightly determined that the Green was a major branch of the Colorado River that flowed to the Gulf of California. Lewis's fork (of the Colombia River) is now called the Snake River.

John Wyeth's estimation that he traveled 25 miles a day is a little too generous. Using this estimation gives him too great a distance to place him at Pierre's Hole. From Wyeth's account, students will conclude that he was somewhere near the present day Idaho/Oregon border.

Management

1. Have students work in groups of three assuming the following roles: one student reads the journal, another keeps the calendar, and the third tracks the trip on the map. The entire group should be involved in the interpretation of the account.
2. Plan two 45-minute periods for this activity. The first period involves reading the journal and translating its account onto the calendar. The second period engages students in making a scaled string of the journey and determining the route taken on the map.
3. Students will encounter difficulty in interpreting the data from the journal. To determine the itinerary, the account is best worked backwards. Students need to discuss if "about twenty-five miles a day" means every day. There will also be some frustration due to the lack of clarity in the account which results in the need for interpretation. This provides an excellent opportunity for discussion and debate. As there is no right answer, students will need to justify their reasonings.

Procedure

1. Distribute *Diary Daze Information and Journal*, have students read them, highlighting information of time and distance. When students are finished, discuss the *Key Question.*
2. Put the students into small groups and distribute calendars.
3. On the approximate date of occurrence from the 1832 calendar, have the groups develop an itinerary of the expedition by recording the landmarks which were referred to in the account.
4. Using the rate of 25 miles per day, have students calculate the miles traveled by Wyeth's group from Independence to the end of the journey. Direct them to keep a running total of the mileage on the calendar.
5. Using the scale on the map and the students' total calculated distance from Independence to Pierre's Hole, have them determine the distance the scaled journey would be on the map.
6. Have students take a string that is longer than the scaled distance and place a piece of masking tape near one end of the string to represent Independence.
7. The students then cut the string so the length from the masking tape to the end is the scaled total distance of the trip.
8. Using colored markers, direct the students to mark the string at the scale distances from Independence for each of the landmarks. They will need to make a key to match the different-colored marks to the corresponding landmark.
9. To determine their prediction of where Pierre's Hole is, have students place the masking tape on the map at Independence (suburb of Kansas City, Missouri) and then lay the string out along the route they think Wyeth traveled. (The teacher may choose to tell the students about the name changes in rivers before doing this step.)

Discussion

1. What difficulties did you have in completing Wyeth's itinerary of his trip? [need to work backwards from July 6th, lack of clarity]
2. What is lacking that makes it difficult to make a map? [headings, positions]
3. How critical in your prediction is the fact that John Wyeth wrote that they traveled 25 miles a day? [prediction is totally dependent if no adjustment was made]
4. What contradictions did you notice between John Wyeth's account and the maps?

Extension

Pierre's Hole lies just to the west of Grand Teton National Park near the present-day border of Idaho and Wyoming. Using the account of Wyeth and a current map, determine a more reasonable rate of travel.

Curriculum Correlation

Social Studies

1. Have students research the route of the Oregon Trail and compare it to Wyeth's account which probably closely followed the trail as far as South Pass.
2. Have students read the full account by Wyeth to get his total perspective of the outing.
3. Have students research the 1832 rendezvous and the mountain men Wyeth would have met.
4. Have students find some historical maps to compare with current maps. Students should discuss why there are differences and what might have caused them.

Diary Daze 1832

May

Sunday	Monday	Tuesday	Wednesday	Thursday	Friday	Saturday
		1	2	3	4	5
6	7	8	9	10	11	12
13	14	15	16	17	18	19
20	21	22	23	24	25	26
27	28	29	30	31		

June

Sunday	Monday	Tuesday	Wednesday	Thursday	Friday	Saturday
					1	2
3	4	5	6	7	8	9
10	11	12	13	14	15	16
17	18	19	20	21	22	23
24	25	26	27	28	29	30

July

Sunday	Monday	Tuesday	Wednesday	Thursday	Friday	Saturday
1	2	3	4	5	6	7
8	9	10	11	12	13	14

Diary Daze Information and Journal
Wyeth's Journey

Wyeth

In 1832 the American West was located west of the wide Missouri River. Lewis and Clark had explored the Missouri and Columbia River basins from 1804-1806. They brought back a great deal of information about the land bought from France in the recent Louisiana Purchase. Lewis and Clarke had gained a general understanding of the lay of the land, but the maps made from their notes and measurements were missing vast amounts of information. Many inaccuracies resulted from misconceptions and misinformation. Further information about the West was the consolidation of information gathered from fur traders, mountain men, and Indians who were familiar with the area. Their descriptions were informative and provided clues to the land forms of the West, but the accuracy needed for making maps was lacking.

In 1832, a group of twenty-one adventurers led by Nathaniel Wyeth set out overland to Oregon. They had been inspired by the publications of Lewis and Clark, Meckenzie, and Gray which told of the wonders of the western coast of the American continent.

Luckily, the group met up with the experienced trader, William Sublet, who lead them across the frontier. The group crossed the Rocky Mountains and went with Sublet to the largest rendezvous of trappers at Pierre's Hole. Seven members of the group returned to civilization with William Sublet. The rest of the group continued to the Pacific. John Wyeth was one of the seven who returned rather than continue. He wrote of his experience on his return.

The leader, Nathaniel Wyeth, later returned to Boston and led a second expedition to Oregon. He established Fort Hall which became an important stop on the emigrants' trail to Oregon and California.

The following are excerpts from the book written by John Wyeth that provide some insight into the geography of the trip which followed closely the course taken by the emigrants.

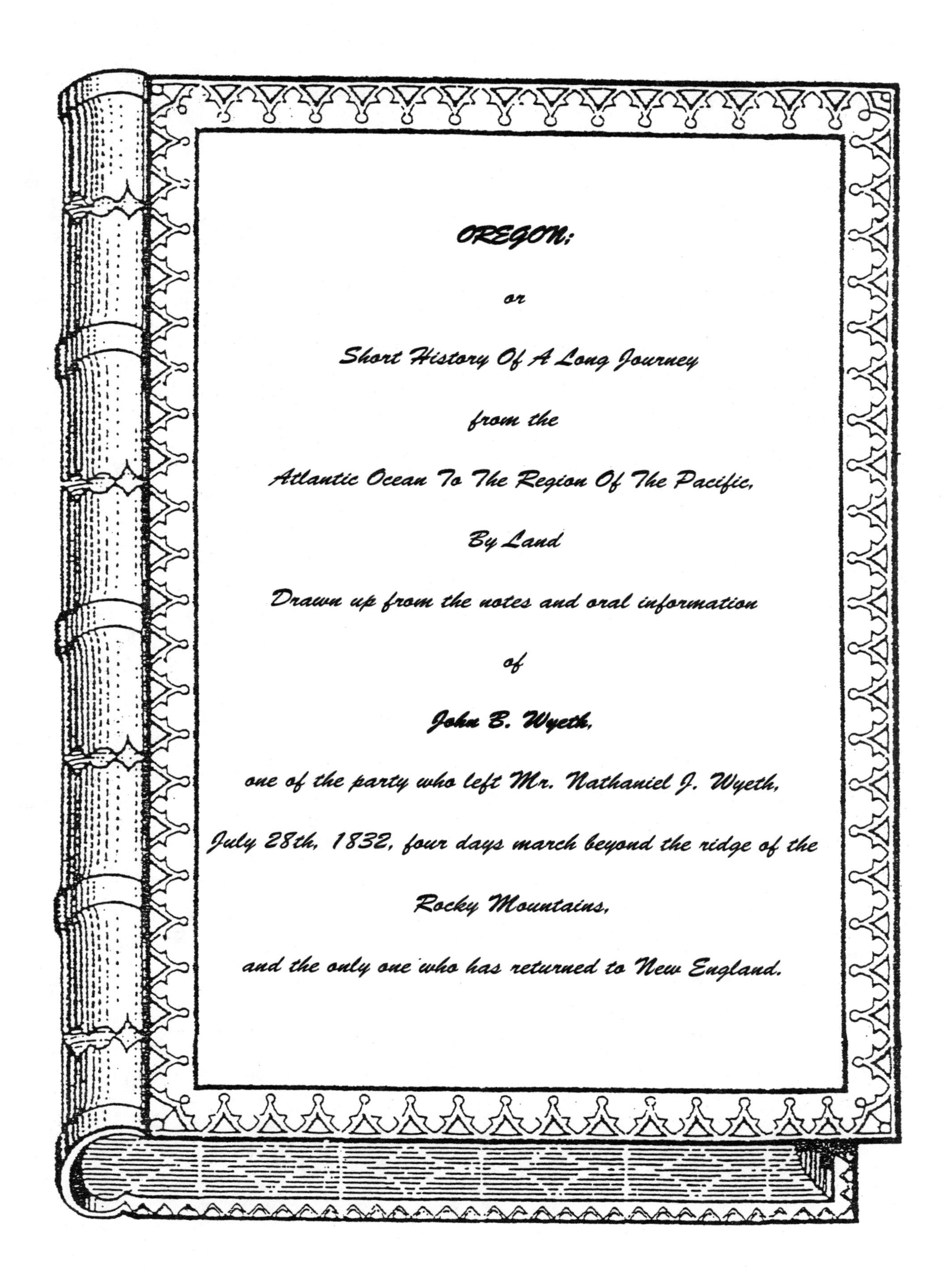

OREGON;

or

Short History Of A Long Journey

from the

Atlantic Ocean To The Region Of The Pacific,

By Land

Drawn up from the notes and oral information

of

John B. Wyeth,

one of the party who left Mr. Nathaniel J. Wyeth,

July 28th, 1832, four days march beyond the ridge of the

Rocky Mountains,

and the only one who has returned to New England.

We arrived at a Town or settlement called Independence. This is the last white settlement on our route to the Oregon....

Just before we started from this place, a company of sixty-two in number arrived from St. Louis, under the command of William Sublet, Esq., an experienced Indian Trader, bound, like ourselves, to the American Alps, the Rocky Mountains, and we joined company with him, and it was very lucky we did.

From this place we travelled about twenty-five miles a day.

Nothing occurred worth recording, till we arrived at the first Indian settlement, which was about seventy miles from Independence.

We travelled on about a hundred miles farther, when we came to a large prairie, which name the French have given to extensive tracts of land mostly level, destitute of trees, and covered with tall, coarse grass.

In sixteen days more we reached the River La Platte, the water of which is foul and muddy. We were nine days passing this dreary prairie. We were seven and twenty days winding our way along the borders of the La Platte, which river we could not leave on account of the scarcity of water in the dry and comfortless plains.

We travelled six days on the south branch of the La Platte, and then crossed over to the north branch of it, we travelled eighteen days.

From the North branch we crossed over to what was called Sweet-water Creek.

We came to a huge rock in the shape of a bowl upside down. It bore the name of Independence, from it is said, being the resting-place of Lewis and Clark on the 4th of July; ...

We had now certainly begun our ascent to those lofty regions, previous to which we had to pass the chief branch of the river La Platte; ...

About the first of July we crossed the highest part or ridge of the mountains.

For the last five days we have had to travel on the Colorado of the West, which is a very long river, and empties into the gulph of California.

On the 4th of July, 1832 we arrived at Lewis's fork, one of the largest rivers in these rocky mountains. It took us all day to cross it.

From the north fork of Lewis's river we passed on to an eminence called Teton mountain, where we spent the night.

On the 5th of July we started afresh rather low-spirited.

Captain Sublet's grand rendezvous, or Head Quarters, was about twelve miles from our encampment.

It was now the 6th of July, 1832 being sixty-four days since we left the settlements of the white people.

This information gives a description of the geography of the emigrants' trail that would be followed to Oregon. With a map and the author's description, can you determine where he ended up?

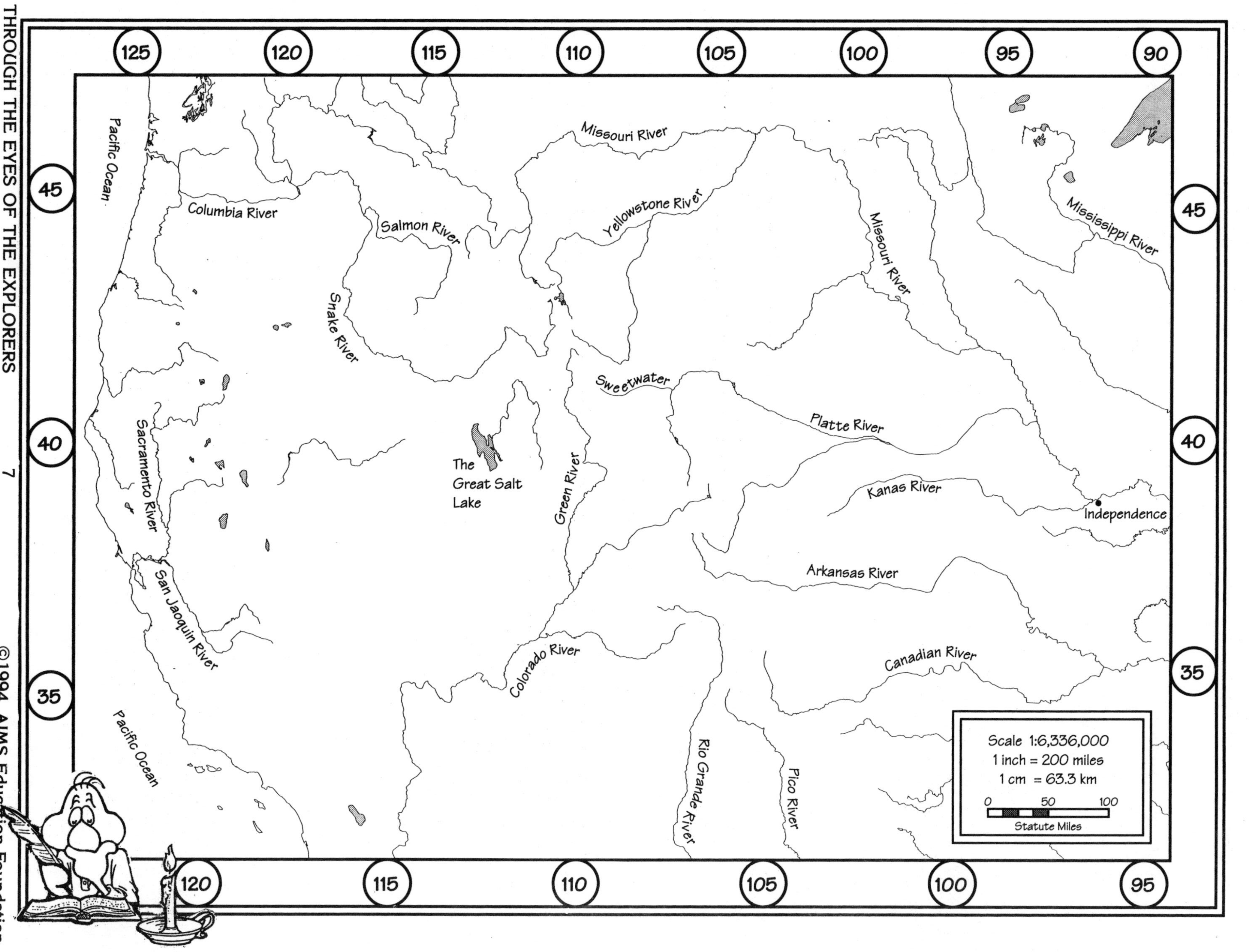

125
120
115
110
105
100
95
90
45
40
35
Pacific Ocean
Columbia River
Salmon River
Missouri River
Yellowstone River
Missouri River
Mississippi River
Snake River
Sweetwater
Platte River
The Great Salt Lake
Green River
Sacramento River
Kanas River
Independence
Arkansas River
San Jaoquin River
Colorado River
Canadian River
Pacific Ocean
Rio Grande River
Pico River
Scale 1:6,336,000
1 inch = 200 miles
1 cm = 63.3 km
0
50
100
Statute Miles

MAP KEY

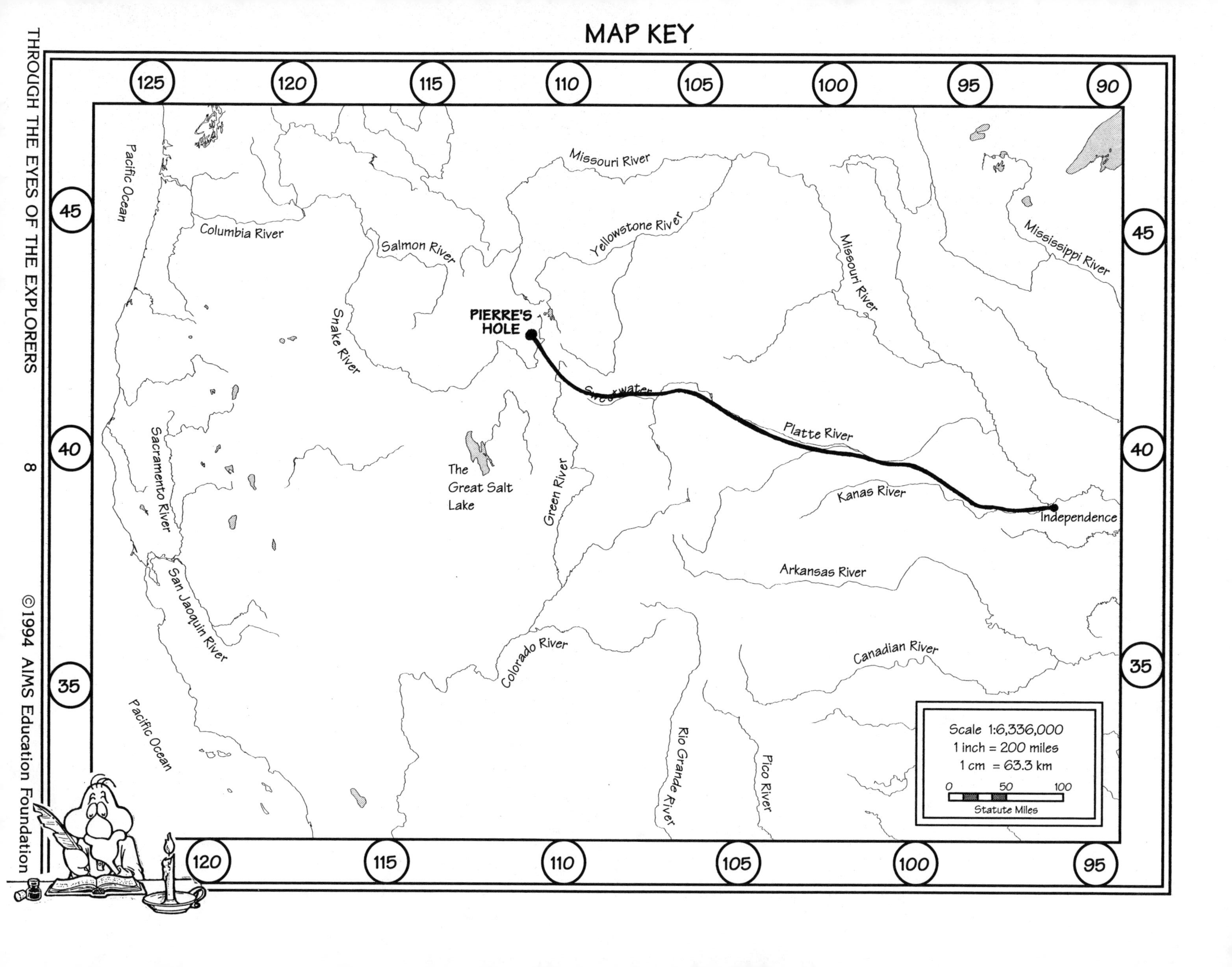

125
120
115
110
105
100
95
90
45
40
35
Pacific Ocean
Columbia River
Salmon River
Missouri River
Yellowstone River
Missouri River
Mississippi River
Snake River
PIERRE'S HOLE
Platte River
The Great Salt Lake
Green River
Sacramento River
Kanas River
Independence
Arkansas River
San Jaoquin River
Colorado River
Canadian River
Pacific Ocean
Rio Grande River
Pico River
Scale 1:6,336,000
1 inch = 200 miles
1 cm = 63.3 km
0
50
100
Statute Miles
45
40
35
120
115
110
105
100
95

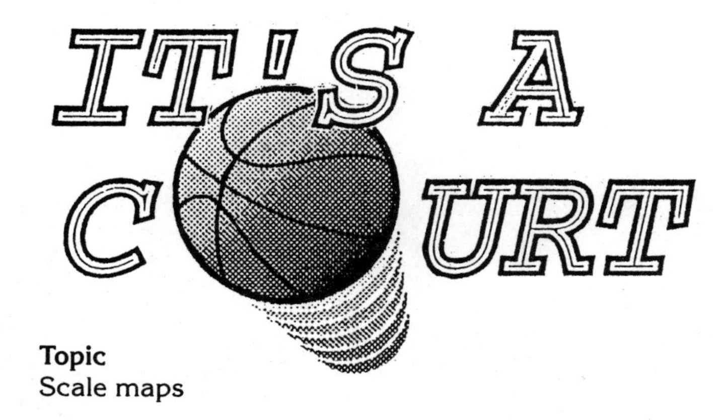

Topic
Scale maps

Key Question
How would you make a map of the basketball court?

Focus
In this activity students measure a basketball court and make a scale drawing of it.

Guiding Documents
NCTM Standards
- *Understand and apply reasoning processes, with special attention to spatial reasoning and reasoning with proportions and graphs*
- *Explore problems and describe results using graphical, numerical, physical, algebraic, and verbal mathematical models or representations*

Project 2061 Benchmarks
- *The scale chosen for a graph or drawing makes a big difference in how useful it is.*
- *Mathematical ideas can be represented concretely, graphically, and symbolically.*
- *Scale drawings show shapes and compare locations of things very different in size.*

Math
Measurement
  length
Proportional reasoning

Integrated Processes
Observing
Collecting and recording data
Interpreting data
Applying

Materials
30 meter tape measure
Rulers
Drawing compasses

Background Information
If you cannot measure an actual basketball court, the dimensions of a regulation high-school court are as follows:

Length: 29 m
Half court length: 14.5 m
Width: 15.4 m
Diameter of restraining circle: 3.7 m
Diameter of center circle: 1.2 m
Width of free throw lane: 3.7 m
Length of free throw lane: 5.8 m

Management
1. This activity is done best in small groups. If only one tape measure is available, it may be done as a whole class activity which makes the measurement of the court faster.
2. Encourage students to round their measurements to the nearest 0.1 meter.
3. The scale has been left blank on the sketch sheet to allow students to use whatever scale they feel is appropriate. If students are to do the drawing on the student sheet provided, 0.8 cm for the scale may be inserted before copying it for the students.
4. The drawing space provided on the student sheet has dark grid lines every 0.8 cm. This allows a regulation court to be drawn on the sheet at a scale of 1 meter to 0.8 of a centimeter. This grid arrangement allows students to draw the court without doing the mathematical conversion. They merely use the grid to count each bold line as a meter. Very few students realize this situation until after they have completed the drawing. This, however, does provide a useful insight for future drawings.
5. The teacher may find it appropriate to have students generate their own sketch on which to record dimensions. Plain paper or graph paper may be used for the drawing. Graph paper helps students keep things at right angles and provides more accurate maps.

Procedure
1. Discuss *Key Question* with students.
2. Distribute materials.
3. Have students make needed measurements of basketball court and record them on the student sheet.
4. Direct them to calculate scaled measurements using proportions and record them on the student sheet.
5. Have students make scale drawings of the basketball court.

6. Have them make measurements from their drawings and determine corresponding measurements on the actual court.
7. The students should then return to the court to check the accuracy of their maps by checking their solutions to *Procedure 6.*

Discussion
1. Using your drawing and a ruler determine how far a player would have to throw a pass from one corner of the court to a teammate on the center line on the opposite side. [21.2 m on regulation court]
2. How long would a pass be if thrown diagonally from one end line to another? [32.8 m]
3. How could you have used the grid on the drawing space to help you make the drawing? [count bold grid as meters instead of scaling]
4. Why did you need to measure only to the nearest 0.1 meters? [scaled distance is so small]

Extension

Make scale maps of the classroom, library, cafeteria and other rooms in the school; include doors and fixtures.

To make a map of the basketball court, measure all the needed dimensions to the nearest 0.1 meter. Record your measurements in the boxes on the sketch below.

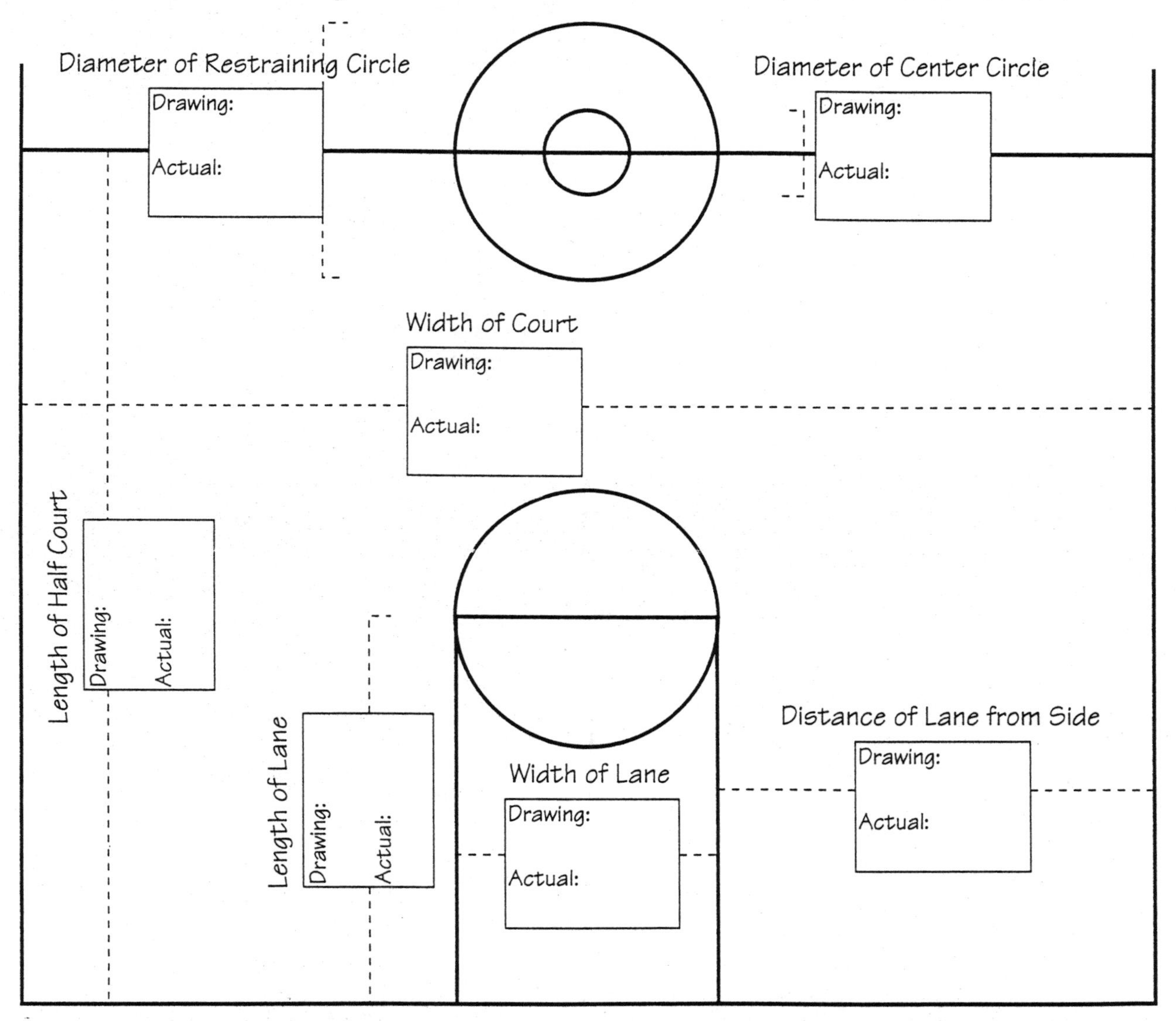

Change each of the actual measures into a scaled measure for the drawing. Put the measure for each dimension above in the actual measure position and complete the proportion to find the scaled measure.

$$\frac{\textbf{Drawing}}{\textbf{Actual}} \qquad \frac{\_\_\_\_\ \textbf{cm}}{\textbf{1 meter}} = \frac{\text{Scaled Measure}}{\text{Actual Measure}}$$

Use the scaled measures to complete your drawing of the basketball court. Use your ruler to make accurate measures and a compass to make a neat drawing.

It's a Court Case

Drawn by:

Date:

Scale:

"Honey, I shrunk the..."

Topic
Scale drawing

Key Question
How could we shrink this room so it would be proportionally the same size for a toy action figure?

Focus
Students will draw a map of the classroom or another room to the scale of a toy action figure.

Guiding Documents
NCTM Standards

- *Understand and apply reasoning processes, with special attention to spatial reasoning and reasoning with proportions and graphs*
- *Understand and apply ratios, proportions, and percents in a wide variety of situations*

Project 2061 Benchmarks

- *The scale chosen for a graph or drawing makes a big difference in how useful it is.*
- *Estimate distances and travel times from maps and the actual size of objects from scale drawings.*

Math
Measurement
Proportional reasoning

Integrated Processes
Observing
Collecting and recording data
Interpreting data

Materials
Toy action figures
30-meter tape measures
Meter sticks
Rulers
Butcher paper

Background Information
To make a scale drawing proportionate for an action figure, a typical real-world height needs to be compared to the action-figure's height. Assume a typical adult stands 175 centimeters high. If an action figure had a height of 12 centimeters the scale would be 175 cm : 12 cm. This scale is simplified to 14.6 cm: 1 cm ($175 \div 12 \approx 14.6$). This scale tells us that measurements in the real world would be 14.6 times bigger than what they need to be for the action figure. The real-world measurements are converted by dividing them by 14.6 to determine the proportionate size for the action figure.

A scale drawing of a room usually includes only the permanent fixtures such as cabinets, counters, sinks, book cases, closets, and radiators. The placement of doors and windows is also indicated on a scale drawing. Furniture is not included.

When making a scale drawing, it is not only important to get the dimension of the object, but its placement in relationship to the room. A door may be 90 centimeters wide, but its location along the wall is also important. It is helpful to draw a rough sketch of the room on scratch paper. As measurements are taken, they can be recorded on the sketch and any missing measurements will become apparent.

Management

1. Groups of three work well for this activity, two to measure and one to record.
2. Generally it takes two 45-minute periods, to complete this activity. The first period is spent gathering measurements, and the second is spent making the scale drawing.
3. Prior to this activity ask students to find a toy action figure at home and bring it to school for the project. This will save the teacher time and expense.
4. Make sure all measurements are made in centimeters.
5. If it is appropriate, the teacher may choose not to give the chart to the students, but instead have them construct one of their own design. Some students may find it more meaningful to record the scaled length of the *Miniature World* on their drawing by the corresponding *Real World* length.

Procedure

1. Have students bring action figures to class.
2. Discuss the *Key Question.*
3. Have students make a sketch of the classroom with its fixtures.
4. Divide the students into groups and distribute the materials.

5. Have students measure and record the necessary measurements of the room.
6. Direct them to measure and record the heights of their action figures and determine what scale to use.
7. Have students use the scale to convert the real-world measures to the miniature-world equivalents, and record them on the chart.
8. Distribute butcher paper and have the students use the miniature-world measures to make a scale drawing of the room.

Discussion
1. Did all the scale drawings of the room come out the same size? (They won't if students are using different-sized action figures.)
2. Why were the drawings different sizes if they represent the same room? [Different scales were used for different action figures because the sizes of the action figures were different.]
3. How might we compare the different-size drawings to see if they are proportionally the same size? [Compare the same dimensions using the corresponding action figures. If a dimension is two action figures long on its drawing, the same dimension on another drawing should be two of its action figures long.)

Home Links
1. Have students make a map of a room in their home to the scale of their action figure.
2. Have students construct a model of a room in their home to the scale of the action figure. Have them include the furniture. They may want to paint and decorate it.

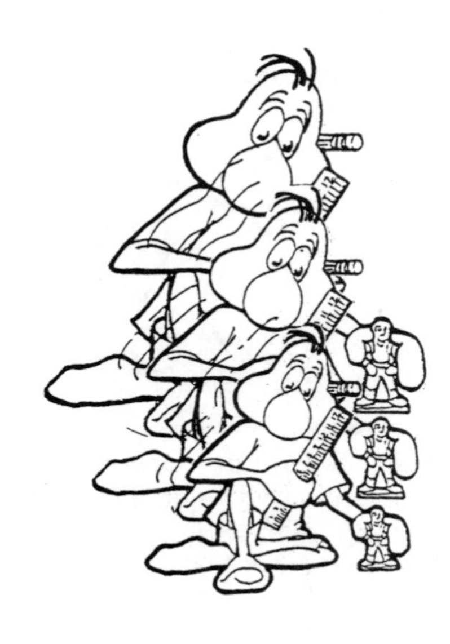

"Honey, I shrunk the..."

Measure the height of your action figure and record on the chart below. Calculate the scale needed to make the drawing proportionate to the action figure. Record each dimension you measured in the room under *Real World* and convert it to the scaled length of the *Minature World* of your action figure. Use the *Minature World* dimensions to make your drawing.

	Real World : Miniature World
Action Figure Height	175 cm : _____ cm
Unit Scale for Drawing	_____ cm : 1 cm

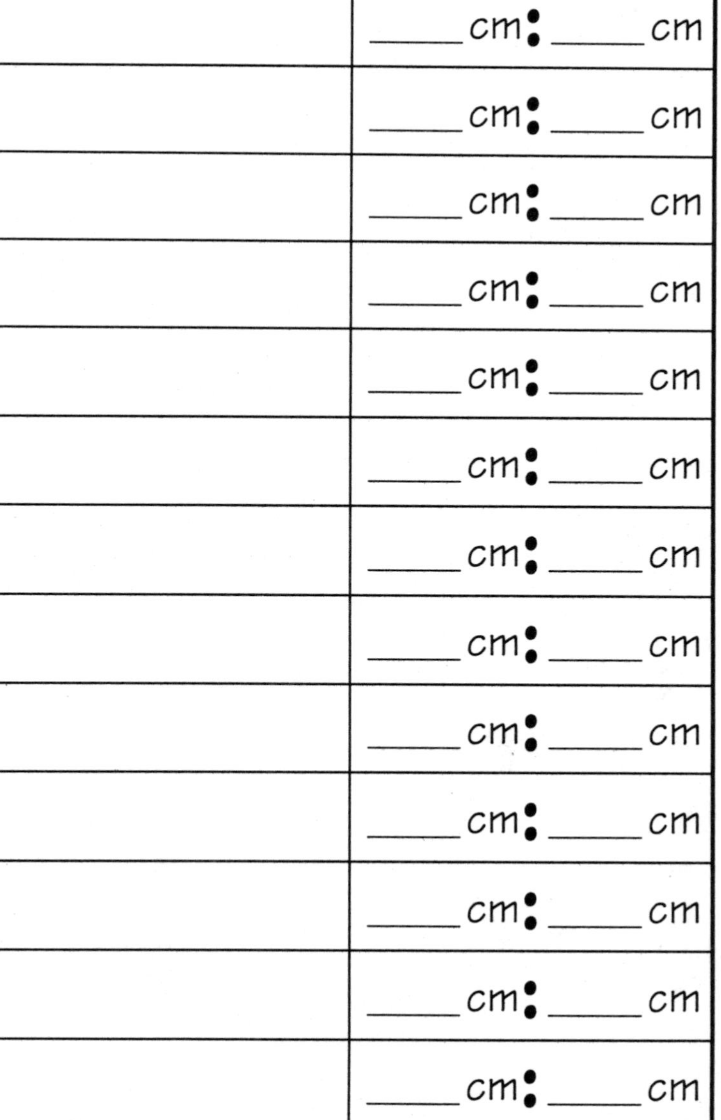

Dimension Description	Room Measurements
	_____ cm : _____ cm
	_____ cm : _____ cm
	_____ cm : _____ cm
	_____ cm : _____ cm
	_____ cm : _____ cm
	_____ cm : _____ cm
	_____ cm : _____ cm
	_____ cm : _____ cm
	_____ cm : _____ cm
	_____ cm : _____ cm
	_____ cm : _____ cm
	_____ cm : _____ cm
	_____ cm : _____ cm

A PACE RACE

Topic
Using non-standard units of measure

Key Question
How many paces do you think you would take to walk the 2,000 miles from Independence, Missouri to California or Oregon?

Focus
Students will determine the length of their paces and use them to calculate how many steps they would need to take to walk the Oregon Trail.

Guiding Documents
NCTM Standards

- *Extend their understanding of the process of measurement*
- *Generalize solutions and strategies to new problem situations*
- *Develop and apply a variety of strategies to solve problems, with emphasis on multistep and nonroutine problems*
- *Describe and represent relationships with tables, graphs, and rules*
- *Estimate, make, and use measurements to describe and compare phenomena.*

Project 2061 Benchmarks

- *Numbers can be written in different forms, depending on how they are being used. How fractions or decimals based on measured quantities should be written depends on how precise the measurements are and how precise an answer is needed.*
- *The graphic display of numbers may help to show patterns such as trends, varying rates of change, gaps, or clusters. Such patterns sometimes can be used to make predictions about the phenomena being graphed.*
- *Organize information in simple tables and graphs and identify relationships they reveal.*

Math
Measuring
  length
Patterns and functions
Graphing
  slope
Estimating
Proportional reasoning
Using rational numbers
  fractions
  decimals

Integrated Processes
Observing
Predicting
Collecting and recording data
Interpreting data
Applying

Materials
100 ft. tape measure
Calculators

Background Information
In the middle of the 19th century, the Oregon Trail and its branch, the California Trail, were the main overland routes to the west coast. Many of the settlers walked the trip from Independence, Missouri to California. Even those who started on wagons, horses, or mules found themselves walking as wagons broke and animals became either food or beasts of burden. The trip, depending on which variation was followed, was about 2,000 miles.

A *pace* is defined as a unit of linear measure equal to the length of one step. In this activity, *pace* is further defined as the distance from the toe on one foot to the toe on the other foot when the feet are spread the distance of one step.

Management
1. Historically, people in the United States have used the customary system of measurement which uses feet and miles. Since this activity is related to a historical event, this customary system is used. If metric units are desired, have students use a meter tape and change the *Key Question* to include a distance of 3,200 kilometers.
2. Before doing the activity, find an area where the length of the 100-foot tape can be laid out with ample room to walk. Establish a starting line at the beginning of the tape measure.
3. If only one tape measure is available, you will need to divide the class into groups of four to do the pacing. Two students can pace, while the other two students in the group record; then they can switch roles so all students will know the lengths of their paces.

Procedure

1. Discuss the *Key Question* and how students might go about getting a good estimate.
2. Distribute data sheets and have students write in their predictions of the number of steps they think they will need to take to walk the 2,000 mile journey.
3. To have students gather data for 5, 10, 15...paces, inform them that they will have to start with their toes touching the 0-inch end of the measuring tape for **each set** of paces.
4. Instruct students to convert their findings of feet and inches into decimal feet. For example, 15'9" becomes 15.75'.
5. Have students find the differences between adjacent pacing lengths and record them in the spaces to the right of the chart.
6. Ask students what patterns they see in the chart. Focus their attention on the *Difference in Distance* column. [Distance gets greater by almost the same amount for each five paces.]
7. Ask students how they might decide how far they go with each five paces. [average the differences]
8. Have students find the *Average Difference* and record it.
9. Ask students how they could predict the distance they would go in 50 paces by using the information on the chart. [Add the distance of 20 and 30 paces. Add four more average differences to the distance of 30 because there are four more groups of five paces. Multiply the average difference for five paces by 10]
10. Instruct students to make a graph of their data.
11. Ask them what patterns they see in the graph. [The data form a line that is close to straight.]
12. As a class, discuss the similarity in the *Differences* on the chart and the line of the graph. [The section with the greatest difference is the steepest. The section with the least difference is less steep.]
13. Have students draw a line from the origin to the last point (30 paces).
14. Ask students to look at the graph and determine how many feet the line drawn in *Procedure 13* goes up for each five paces it goes to the right. How is this line similar to the *Average Difference*? [The line goes up the *Average Difference* each five paces.]
15. Have students explain how they could use their graphs to predict how far they would go in 22 paces and 35 paces. [For 22 paces, they could use the line and interpolate the distance traveled in 22 paces. By extending the line, they could extrapolate the distance traveled in 35 paces. (You might want to have them check their predictions by pacing.)]
16. Ask students to determine and record the length of one pace.
17. Discuss how the length of one pace shows up on the graph. [It is the slope of the line. It goes up the length of one pace for each pace it goes to the right.]
18. Have students write a function relating the distance they traveled to the number of paces they walked. [Distance Traveled = Length of One Pace X Number of Paces]
19. Instruct students to determine how many steps they would have to take to make the journey of 2,000 miles.

Discussion

1. How close was your prediction to your calculated estimate?
2. Is your calculation a realistic estimate of the number of steps that you would take on the journey? What things might make the actual count different? [trail conditions, grade, weather, personal physical condition]
3. How accurate is the measurement of your pace? Explain.

Extensions

1. Have students use their stride (paces) to measure objects. Have them check their measurements with a measuring tape or ruler.
2. Have students research the amount of time it took the pioneers to make the journey and determine how many steps per day (using the measure of their pace) they would have to take.
3. Gather the average distance per pace from each student in the class. Make a statistical study to determine what a typical student's stride is.

A discussion of unit conversion as a format by which to solve this problem might be appropriate for some students. An example of this format using a pace of 1.5 ft:

a. Set up the problem
(2,000 mi x 5,280 ft/mi) ÷ 1.5 ft/pace

b. Separate the numbers and units
(2,000 x 5,280) ÷ 1.5) (mi x ft/mi ÷ ft/pace)

c. Calculate numbers, invert division to multiplication in units
(10,560,000 ÷ 1.5) (mi x ft/mi x pace/ft)

d. Cross out units and calculate
7,040,000 paces

A PACE RACE

The pioneers traveled 2,000 miles from Independence, Missouri to California and Oregon. Many of those settlers had to walk the entire way. How many steps (paces) do you think it would take for you to walk this same distance?

Your Prediction of Paces [ ]

Complete the chart and the graph to determine how long your pace is.

Paces	Distance Feet	Distance Inches	Feet as a Decimal
5		/12	
10		/12	
15		/12	
20		/12	
25		/12	
30		/12	

Difference in Distance

[ ]
[ ]
[ ]
[ ]
[ ]

Average Difference [ ]

[ ] **Feet ≈ 5 Paces**

[ ] **Feet ≈ 1 pace**

How many steps would it take you to cover the 2000 miles?

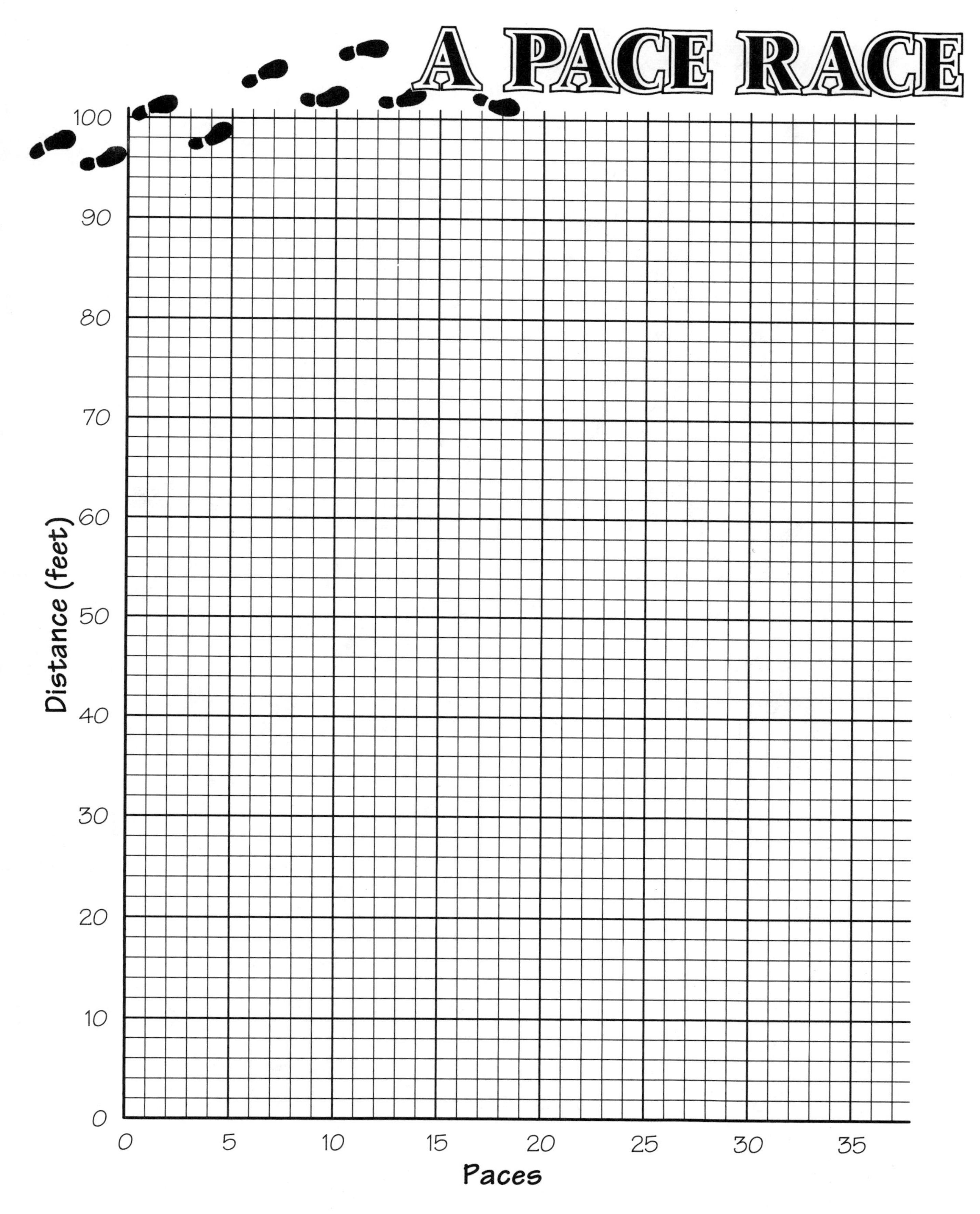

Write a function that will tell how many feet you have traveled, when you substitute how many paces you have taken.

The Yard

Topic
Measuring and mapping

Key Question
If you do not have something with which to measure the school grounds, how could you make a scale map of the area?

Focus
Students will construct a map of the school grounds using pacing as their measuring method.

Guiding Documents
NCTM Standards

- *Extend their understanding of the process of measurement*
- *Use problem-solving approaches to investigate and understand mathematical content*

Project 2061 Benchmarks

- *Estimate distances and travel times from maps and the actual size of objects from scale drawings.*
- *The scale chosen for a graph or drawing makes a big difference in how useful it is.*

Math
Measurement
Proportional reasoning
Scale drawing

Integrated Processes
Observing
Collecting and recording data
Interpreting data
Modeling

Materials
Graph paper
Rulers

Background Information
Many maps of the explored areas were based on data gathered by surveyors. The coastlines and colonized areas were mapped relatively accurately because of this, but many early maps were based on speculation. To complete the maps of the unexplored areas, cartographers depended on the accounts made by natives and those who went on expeditions. This information usually used periods of travel time and directions in rough compass headings to convey distances. An example might be, "We continued to paddle north up the river for four days." To use this information for a map, the cartographer would have to decide how far an explorer could paddle a canoe up river every day. Using this estimation, this portion of the river would then be drawn to scale on a map.

Pacing provides a similar experience for the students. A pace gives an approximation of distance. The paces can be converted into feet, scaled and drawn on a map. Although this map may not be precise, it provides the general size and configuration of the area.

Management

1. This activity is written to be done after doing the activity *A Pace Race*. If that activity has not been done, students will need to determine the length of paces.
2. This activity works well with students in groups of three. One student will pace while another records the paces on a sketch of the school. The third student records the paces on the conversion chart. If group management is a problem, the activity can be done as a whole-class activity with different individuals doing the pacing for different measurements. All students can record the measurements on the conversion chart. Each individual will need to know their pace to foot ratios.
3. Allow at least two periods for this activity: one to gather the school's measurements of the school by pacing, and the second to convert the paces to standard measures and make the map.
4. Before doing the activity, the teacher should choose an appropriate part of the school to map. If the building is rather large and sprawling, the students might map only one wing or area.
5. When drawing the map, there are two general methods chosen by students for determining scale.

   a. One is to scale the map to paces making each square length (1/4") equal to a certain number of paces. This is the simplest because it takes no conversion. However, when using this method the foot equivalents need to be recorded on the map along each measurement.

b. The second method is more conventional. Students convert the foot distances to an inch scale. The students divide the greatest foot measurement by 10, (the height of the grid in inches.) The quotient is the minimum feet-per-inch scale that will fit on the grid. Students will want to round this quotient up to a whole number. Encourage students to round the quotient to a scale that will simplify their work. Choosing a multiple of four allows students to make use of the 1/4" marks.

6. By gluing or taping grids together, students can make maps of larger areas or use larger scales.

Procedure

1. Have students discuss the *Key Question*.
2. Distribute record sheets and scratch paper for sketches.
3. Take students to area that is to be mapped and make sure students have an idea of the area's layout.
4. Have students make a sketch of the area's layout as an overhead view.
5. Direct the students to pace the lengths of the necessary measurements, record them on the record sheet, and sketch.
6. Check to be sure that they record the feet-to-pace ratio of the students who paced each dimension.
7. Have them calculate and record the length of each measurement to the nearest foot.
8. Distribute the grid paper. Have students choose a scale that is appropriate and record it on the top of the *Scaled Distance* column on the record sheet.
9. Have them convert each foot measurement to its scaled distance and record it.
10. Direct the students to use the data from their record sheet and sketch to make a scale drawing of the area on the grid paper.

Discussion

1. How does your map compare to other students' maps?
2. What might have caused these differences?
3. What uses might there be for this kind of map? [landscape, site planning]
4. For what uses would this type of map be inappropriate? [construction, legal documents such as titles]

Extension

Have students return to the mapped area with tape measures or trundle wheels and their maps. Have them check the accuracy of their maps and discuss how they might make them more accurate in the future.

Curriculum Correlation

Social Studies

Have students make a map from the information provided in *Diary Daze*.

Home Link

Have students make a site plan of their home, apartment, or local park by pacing.

The Yard

Measurement Description	Paces X	Ratio (ft/pace) =	Feet

Scaled Distance Scale =___ft/ 1 in

PACING
The Yard

Topic
Longitude and latitude

Key Question
How could you use the data recorded from Fremont's expeditions to make a map of the Oregon Trail?

Focus
Students will reconstruct a map of the Oregon Trail using longitude and latitude sightings taken during Fremont's expeditions.

Guiding Documents
NCTM Standards

- *Understand and appreciate the need for numbers beyond the whole numbers*
- *Compute with whole numbers, fractions, decimals, integers, and rational numbers*

Benchmarks 2061

- *Important contributions to the advancement of science, mathematics, and technology have been made by different kinds of people, in different cultures, at different times.*
- *Numbers can be written in different forms, depending on how they are being used. How fractions or decimals based on measured quantities should be written depends on how precise the measurements are and how precise an answer is needed.*
- *It takes two numbers to locate a point on a map or any other flat surface. The numbers may be two perpendicular distances from a point, or an angle and a distance from a point.*
- *Estimate distances and travel times from maps and the actual size of objects from scale drawings.*

Math
Graphing
Scale measure

Science
Earth science

Integrated Processes
Interpreting data
Comparing and contrasting

Materials
Student Information sheets
Latitude and longitude grid
Current map with trail

Background Information

In the early 1840's the emigration of pioneers to the area of Oregon and California was increasing greatly. In order to have a clearer understanding of the area, Capt. John Charles Fremont of the U.S. Army was assigned an expedition to gather information of the emigrant road and its environs, as well as the west.

Fremont led two expeditions along parts of the emigrant road. The first in 1842 was to the Wind River Range in present-day Wyoming. The second from 1843-4 was to the Willamette Valley in Oregon via the emigrant road, and returning via Alta California (present-day California) and then through what would become the western states of Nevada, Utah, Colorado, and Kansas. Fremont made a third expedition in 1845-6 to gather more information about Alta California and the possibility of a more southern route of travel and commerce.

Fremont was trained as a surveyor. He made measurements of longitude and latitude by taking measurements of the sun and stars, by observing the moons of Jupiter, and by using a chronometer (an accurate timepiece). Using these measurements, Fremont could plot the location of primary landmarks on a grid of longitude and latitude. This provided an accurate map of his travels. Distances between landmarks could be determined by scaling the distances on the map.

Fremont did not make a continuous trip down the emigrant road. He took many alternate routes and side trips. In this activity, to produce a map of the emigrant road, students should only use the selected measurements taken along that trail.

Fremont's chronometer was broken on the return trip of the 1842 expedition. From that point on, only latitude could be determined. The points missing longitude measurements happened between two lengths of the trail measured on the outward journey. In reconstructing the map, students will need to make an approximation of what the longitude at each point is. They will need to assume that the trail followed a straight line between the two measured segments. This will closely approximate the trail.

Degrees of longitude are measured east and west. around the earth. There are 90 degrees of latitude from the equator to the North Pole. Each of these degrees is divided into 60 sections called minutes. Each minute is divided into sixty seconds. Fremont took very accurate measurements recording them to the nearest second, or the nearest 3600th of a degree.

Management

1. This activity works well in groups of three students. Each student plots a segment of the trail. One student plots the 1842 measurements. The second student plots the August, 1843 measurements. The third student plots the September through November, 1843 measurements. The students in each group then combine their three sections to make one map.
2. Students will need to be familiar with longitude and latitude and how to plot them on a grid before beginning this activity.
3. Fremont's measurements were to the nearest second, which is much too precise for the scale of this map. Students will be able to generate a good map by estimating to the nearest quarter of a degree. This can be done by having students write the minutes as a fraction after having them use the seconds to round to the nearest minute. Then they convert the fraction to a decimal.
   Example: 37 degrees, 42 minutes('), 32 seconds(")
   (round to nearest minute) 37°43'
   (convert to fraction) 37 43/60°
   convert fraction to decimal) 37.7267°
   (convert to nearest quarter degree) 37.75°
   Students will soon generalize that 15 minutes is 1/4 of a degree, 30 minutes is 1/2 a degree, and 45 minutes is 3/4 of a degree.
4. The teacher may find it convenient to copy the current map with the trail onto overhead transparency film to be used by students as an overlay to check their maps. If not using the transparency film, students will need to place the current maps over their constructed maps and hold them up to a light, such as a window, to check them.

Procedure

1. Distribute the *Overland Mapping Record* and discuss the *Key Question.*
2. If students are not familiar with longitude, latitudes, degrees, minutes, and seconds, provide instruction.
3. Have the students plot the appropriate points on the map and construct the trail by connecting the points as they go. They may want to label points of major landmarks.
4. If students are working in groups, have them combine their segments by tracing each other's segments to get a complete map.
5. Distribute the *Key: Oregon Trail* and have students compare the accuracies of their maps by comparing them.

Discussion

1. How closely did your map match the Oregon Trail?
2. What might have caused your errors? [approximation of minutes, approximate points when not all information is given]
3. How did Capt. Fremont determine his longitude and latitude? *(Have students complete the next 3 activities.)*

Extension

Fremont did not travel the Oregon Trail in a continuous fashion. Have students use the dates given and reconstruct a trip of a "typical emigrant." Assume the emigrant left in early April. Have them construct a calendar and record the dates when landmarks would be seen by the emigrant.

Curriculum Correlation

Social Science

1. Have students do research to find pictures of different landmarks and forts along the Oregon Trail that Fremont mentions and emigrants would have seen.
2. Have students research and read some journal entries and writings of emigrants on the Oregon Trail.

Language Arts

1. Have students write journal entries that are dated and refer to landmarks and events they would have experienced along the trail.

OVERLAND MAPPING Record

In the early 1840's the emigration of pioneers to the area of Oregon and California was increasing greatly. In order to have a clearer understanding of the area, Capt. John Charles Fremont of the U.S. Army was assigned several expeditions to gather information on the emigrant road and its environs, as well as the West.

Fremont

Fremont was exploring the West from 1842 to 1846 on three different expeditions. Fremont took very careful measurements to determine the longitude and latitudes of his positions. He used these measurements to construct a map of the land he traveled through. Following are the measurements he recorded in his report to the United States Senate. Make a map using this information from:

Senate Ex. document 174, 2nd session, 28th Congress: Report of the Exploring Expedition to the Rocky Mountains in the Year 1842, and to Oregon and North California in the Years 1843-'44

Date		Localities	Latitude			Longitude		
			Deg.	min.	sec.	Deg.	min.	sec.
1842								
June	8	Chouteau's lower trading post, Kansas River	39	05	57	94	25	46
	16	Left bank of the Kansas River, seven miles above the ford	39	06	40	95	38	05
	18	Vermillion creek	39	15	19	96	04	07
	19	Cold Springs, near the road to Laramie	39	30	40	96	14	49
	20	Big Blue river	39	45	08	96	32	35
	25	Little Blue river	40	26	50	98	22	12
	26	Right bank of Platte river	40	41	06	98	45	49
	27	Right bank of Platte river	40	39	32	99	05	24
	28	Right bank of Platte river	40	39	51			
	30	Right bank of Platte river	40	39	55	100	05	47
July	2	Junction of North and South forks of the Nebraska or Platte river	41	05	05	100	49	43
		Expedition divides and Fremont's group takes side trip. They reunite on July 16.						
	16	Fort Laramie, near the mouth of Laramie's fork	42	12	10	104	47	43
	23	North fork of Platte river	42	39	25	104	59	59
	24	North fork of Platte river	42	47	40			
	25	North fork of Platte river, Dried Meat camp	42	51	35	105	50	45
	26	North fork of Platte river, noon halt	42	50	08			
	26	North fork of Platte river, mouth of Deer creek	42	52	24	106	08	24
	28	North fork of Platte river, Cache camp	42	50	53	106	38	26
	29	North fork of Platte river, left bank	42	38	01	106	54	32
	30	North fork of Platte river, Goat island	42	33	27	107	13	29
Aug	1	Sweet Water river, one mile below Rock Independence	42	29	56	107	25	23
	4	Sweet Water river	42	32	31	108	30	13
	7	Sweet Water river	42	27	15	109	21	32
		Chronometer breaks. No more longitude measurements can be made.						
Sept	3	North fork of Platte river, right bank	42	01	40			
	4	North fork of Platte river, near Scott's bluffs	41	54	38			
	5	North fork of Platte river, right bank, six miles above Chimney rock	41	43	36			
	8	North fork of Platte river, mouth of Ash creek	41	17	19			
	9	North fork of Platte river, right bank	41	14	30			
	10	North fork of Platte river, Cedar bluffs	41	10	16			

Date		Localities	Latitude			Longitude		
			Deg.	min.	sec.	Deg.	min.	sec.
1843								
Aug	13	Near South pass, on a small affluent to the Sandy fork of Green river	42	19	53			
	13	Small stream, tributary to the Little Sandy river	42	18	08	109	25	55
	14	Little Sandy river	42	15	11			
	15	Green river, left bank	41	53	54	110	05	05
	16	Green river, near old trading post, at point where the road to the Columbia leaves the river.	41	46	54			
	16	Black's fork of Green river	41	37	38	110	10	28
	17	Black's fork	41	29	53	110	25	06
	18	Small stream, tributary to Ham's fork	41	26	08	110	45	58
	19	Muddy river of Ham's fork	41	34	24			
	20	Muddy river	41	39	45			
	21	Bear river	41	53	55			
	21	Bear river	42	03	47	111	10	53
	22	Bear river, above Thomas's fork	42	10	27			
	24	Tullick's fork of Bear river	42	29	05			
	24	Bear river	42	36	56	111	42	08
	25	Beer springs	42	39	57	111	46	00
Sept.	21	Fort Hall	43	01	30	112	29	54
	24	Snake river, above the American falls	42	47	05	112	40	13
	28	Snake river	42	29	57			
	29	Rock creek, of Snake river	42	26	21	114	06	04
	30	Snake river, opposite to the River spring	42	38	44	114	25	04
Oct.	1	Snake river, 2 miles below Fishing falls	42	40	11	114	35	12
	2	Snake river	42	53	40	114	53	04
	3	Ford where road crosses the Snake river	42	55	58	115	04	46
	7	Big Wood river, or Rivière Boisée	43	35	21	115	54	46
	8	Big Wood river, or Rivière Boisée	43	40	53	116	22	40
	10	Fort Boisée	43	49	22	116	47	03
	12	Snake river, below Birch creek	44	17	36	116	56	45
	14	Head water of Burnt river, (Rivière Brulée)	44	37	44	117	09	49
	15	Old bed of Powder river	44	50	32	117	24	21
	16	Powder river	44	59	29	117	29	22
	18	Grand Rond	45	26	47	117	28	26
	19	Blue mountains, east of the summit	45	38	07	117	28	34
	23	Walahwalah river, foot of the mountains	45	53	35	118	00	39
	26	Fort Nez Percé	46	03	46			
	28	Noon halt – left bank of Columbia	45	58	08			
	30	Left bank of the Columbia	45	50	05	119	22	18
	31	Left bank of the Columbia	45	44	23	119	45	09
Nov.	5	Missionary station at the Dalles of the Columbia	45	35	55	120	55	00
	5	Station on hills in rear of the mission	45	35	21	120	53	51
	11	Right bank of the Columbia, 15 miles below the cascades	45	33	09	122	06	15

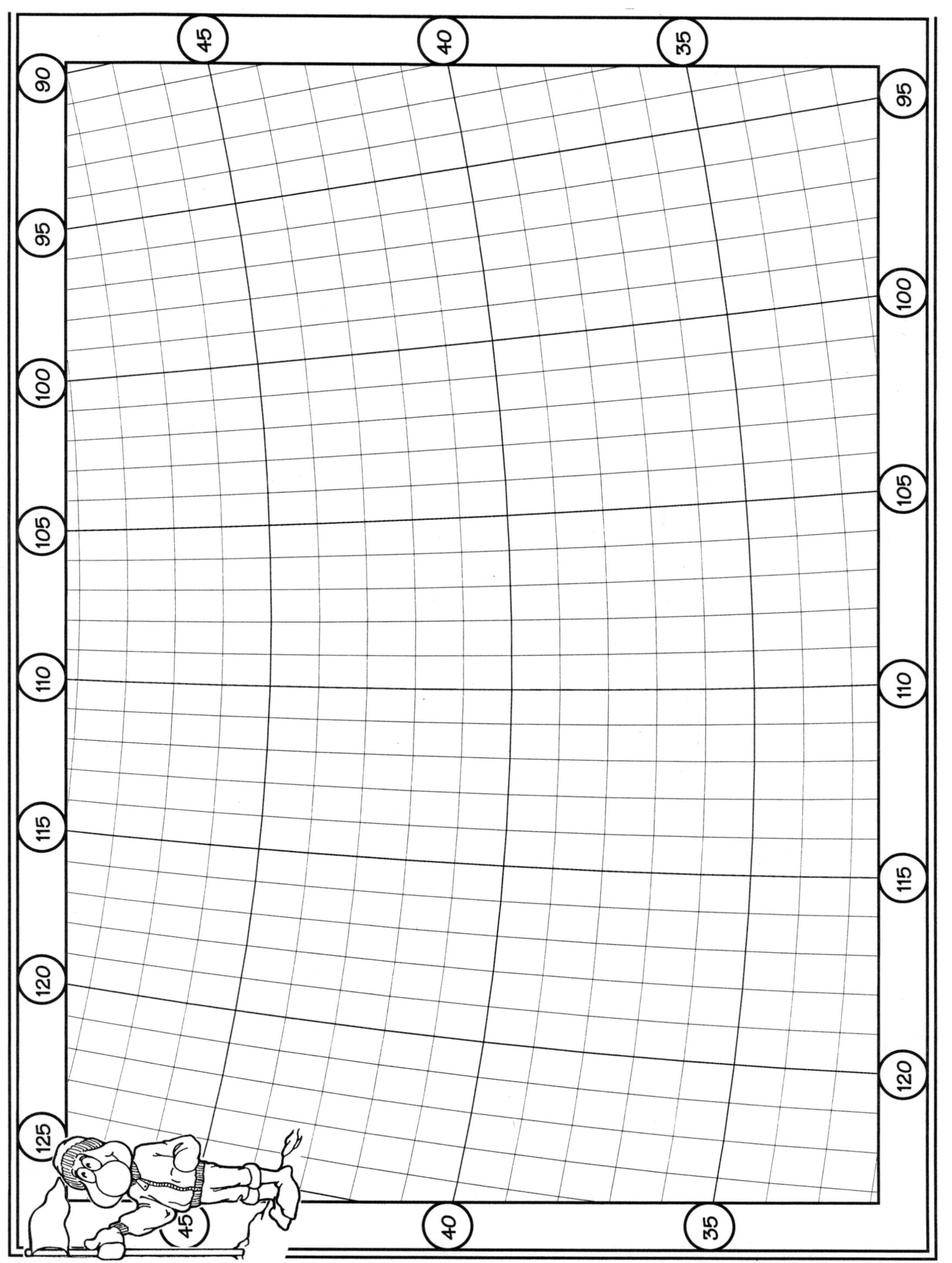

45
40
35
90
95
100
105
110
115
120
125
95
100
105
110
115
120
45
40
35

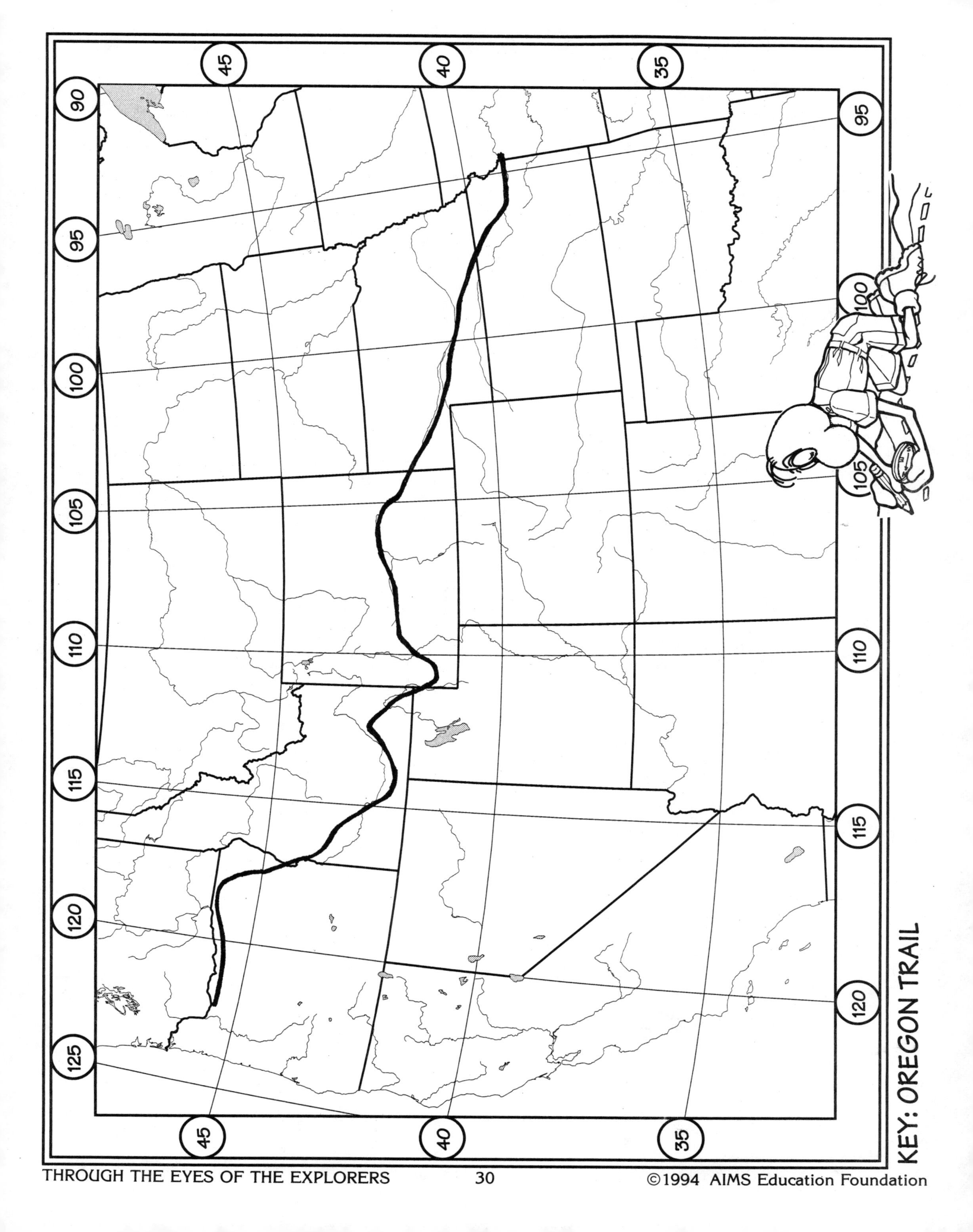

45
40
35
90
95
100
105
110
115
120
125
95
100
105
110
115
120
45
40
35
KEY: OREGON TRAIL

Topic
Orienteering

Key Question
What type of instructions do you need to find a location which was predetermined by someone else?

Focus
Students will use a magnetic compass, tape measure, and a set of instructions to find the location of a treasure.

Guiding Documents
NCTM Standards
- *Develop an appreciation of geometry as a means of describing the physical world*
- *Develop and apply a variety of strategies to solve problems, with emphasis on multistep and nonroutine problems*

Project 2061 Benchmarks
- *Read analog and digital meters on instruments used to make direct measurements of length, volume, weight, elapsed time, rates, and temperature, and choose approximate units for reporting various magnitudes.*

Math
Measurement
  length
  angle
Estimating

Science
Earth's magnetic field
Orienteering

Integrated Processes
Observing
Interpreting Data
Applying

Materials
Magnetic compass
Tape measure or trundle wheel

Background Information
The earth is a large magnet with a north and a south magnetic pole. A magnetized needle floating free will be aligned north and south because of the earth's magnetic field. This idea of a magnetized needle or compass has been known since early Roman times.

By the time of Christopher Columbus, navigators realized that a compass did not always point due north. They were also aware that the amount of discrepancy changed from location to location. This discrepancy (called *declination*) is because the magnetic North Pole is not in the same place as the geographic North Pole. The magnetic North Pole is presently located in northern Canada. Its exact location shifts from time to time, and over geologic time, it has shifted radically.
Navigators now make adjustment for this discrepancy. Below is a map displaying the approximate compass declination for the United States. If you were reading a compass in central California, your compass would show magnetic north as 17° east of geographic north. If you wanted to go geographic north using your compass, you would make this adjustment and would take a compass heading of 343°. With most compasses available in sporting goods stores, you make this adjustment by reorienting the rectangular base with the compass.

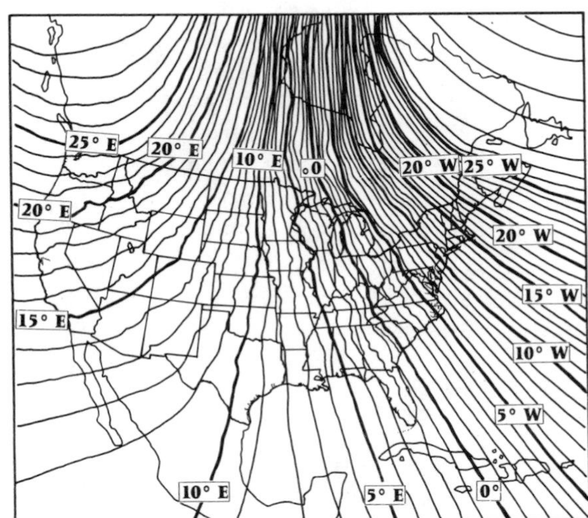

Magnetic declination changes over time. The declination from the map will provide a general idea of the declination of a location. A more accurate record can be obtained by referring to a recent geological survey map, aeronautical map, or by calling a U.S. Department of Transportation Flight Service Station.

Modern navigators do not depend on the magnetic compass alone. A modern jetliner has two compasses, a magnetic and a gyro. Using the most recent magnetic declination information, the navigator identifies true north with the magnetic compass. Using the magnetic compass as a standard, the gyro compass is set to true north. From that point on, the navigator depends on the gyro compass alone for compass settings because the gyro compass is not subject to magnetic declinations.

The composition of the earth's crust also has some affect on the accuracy of the compass. In Minnesota's iron ore mining region, a compass is almost useless due to the influence of the iron in the area.

Management

1. This activity may be done without allowing for compass declination; however, the topic of magnetic declination must be understood if students are to understand how maps are oriented to the earth.
2. Before doing the activity, discuss magnetic declination with your class. Make a transparency of the magnetic declination map and have students determine the declination for their location. Show them how to use a compass and make compensation for magnetic declination.
3. Before doing the activity, find a large open area outside to do the activity. Within that area identify a landmark for each group.
4. Two types of instruction sheets are given, one with the compass headings in degrees and one with the directional headings. (A *heading* is the course or direction something is moving. It is usually expressed as a compass reading in degrees.) Students will have to translate the directions into degrees

North - 0°
Northeast - 45°
East - 90°
Southeast - 135°
South - 180°
Southwest - 225°
West - 270°
Northwest - 315°

The teacher should use whichever instructions are appropriate for the class's needs.

Procedure

1. Discuss the *Key Question*.
2. Distribute materials and instruct students in the use of the magnetic compass.
3. Give each group a list of instructions for finding the treasure.
4. Show each group its assigned landmark.
5. Have students follow directions starting at their landmark.
6. Direct them to record their final position in relation to their assigned landmark, or check them immediately.

Final Position Relative to Landmark

Set A - 10m west
Set B - 16m east
Set C - 14m east
Set D - 11m west

Discussion

1. If you were given another set of instructions, how could you be more accurate? [Answers will vary, but two suggestions are: 1) select a distant landmark to help get straighter lines, 2) have two students use the compass; one makes sure the compass is oriented north while the other makes the directional sighting.]
2. What would have happened if you had not allowed for magnetic declination? [Your final position would have been out of place by the local magnetic declination.]

Extension

Have students develop their own set of instructions for a treasure hunt by following a route out in the field. Have them measure the headings using the compass and the distances with the tape measure and record the measures. Have different groups exchange their sets of instructions to see if they come to the same spot.

Treasure Hunt

Instructions to find treasure

Start at landmark ________

Heading 0°, go 20 meters.
Heading 135°, go 14 meters.
Heading 90°, go 10 meters.
Heading 225°, go 14 meters.
Heading 315°, go 42.4 meters.
Heading 225°, go 28.3 meters.
Heading 135°, go 14 meters.
Heading 90°, go 20 meters.

Here be the **gold!**

A

Instructions to find treasure

Start at landmark ________

Heading 0°, go 7 meters.
Heading 90°, go 7 meters.
Heading 180°, go 14 meters.
Heading 270°, go 14 meters.
Heading 0°, go 21 meters.
Heading 90°, go 21 meters.
Heading 135°, go 9.9 meters.
Heading 225°, go 9.9 meters.

Here be the **gold!**

C

Instructions to find treasure

Start at landmark ________

Heading 225°, go 22.6 meters.
Heading 0°, go 8 meters.
Heading 90°, go 32 meters.
Heading 315°, go 22.6 meters.
Heading 270°, go 8 meters.
Heading 135°, go 33.9 meters.
Heading 45°, go 11.3 meters.
Heading 315°, go 11.3 meters.

Here be the **gold!**

B

Instructions to find treasure

Start at landmark ________

Heading 45°, go 15.6 meters.
Heading 135°, go 15.6 meters.
Heading 225°, go 15.6 meters.
Heading 0°, go 33 meters.
Heading 135°, go 15.6 meters.
Heading 270°, go 22 meters.
Heading 315°, go 15.6 meters.
Heading 180°, go 22 meters.

Here be the **gold!**

D

Instructions to find treasure

Start at landmark ________

Go north 20 meters.
Go southeast 14 meters.
Go east 10 meters.
Go southwest 14 meters.
Go northwest 42.4 meters.
Go southwest 28.3 meters.
Go southeast 14 meters.
Go east 20 meters.

Here be the **gold!**

A

Instructions to find treasure

Start at landmark ________

Go north 7 meters.
Go east 7 meters.
Go south 14 meters.
Go west 14 meters.
Go north 21 meters.
Go east 21 meters.
Go southeast 9.9 meters.
Go southwest 9.9 meters.

Here be the **gold!**

C

Instructions to find treasure

Start at landmark ________

Go southwest 22.6 meters.
Go north 8 meters.
Go east 32 meters.
Go northwest 22.6 meters.
Go west 8 meters.
Go southeast 33.9 meters.
Go northeast 11.3 meters.
Go northwest 11.3 meters.

Here be the **gold!**

B

Instructions to find treasure

Start at landmark ________

Go northeast 15.6 meters.
Go southeast 15.6 meters.
Go southwest 15.6 meters.
Go north 33 meters.
Go southeast 15.6 meters.
Go west 22 meters.
Go northwest 15.6 meters.
Go south 22 meters.

Here be the **gold!**

D

BACKWARD MAPPING

Topic
Earth's magnetic field and compasses

Key Questions
1. How could you make a set of directions for a treasure hunt?
2. How could you make a set of directions without going to the field?

Focus
Students will make a scale map of an orienteering course and then make a set of directions for the course.

Guiding Documents
NCTM Standards
- *Develop common understandings of mathematical ideas, including the role of definitions*
- *See mathematics as an integrated whole*
- *Systematically collect, organize, and describe data*
- *Develop an appreciation of geometry as a means of describing the physical world*

Project 2061 Benchmarks
- *It takes two numbers to locate a point on a map or any other flat surface. The numbers may be two perpendicular distances from a point, or an angle and a distance from a point.*
- *The scale chosen for a graph or a drawing makes a big difference in how useful it is.*

Math
Linear measure
Angular measure
Scale drawings

Science
Earth science
  Earth's magnetic field
  magnetic compass

Integrated Processes
Observing
Collecting and recording data
Interpreting data

Materials
Magnetic compass
Tape measure
Ruler (cm)
Overhead transparencies

Background Information
Refer to the *Background Information* in *Treasure Hunt* for information about the earth's magnetic field and use of a compass.

A scale map of an orienteering course provides a simple way to develop the directions for that course. On a paper with grid lines, the path that will direct the contestant to the goal is drawn. This path can be any combination of straight lines. The positions of north, south, east, and west are written on the corresponding sides of the map. Starting at the beginning of the course, the heading of each line is measured relative to north with a 360° protractor. The grid on the drawing helps to align the protractor. The person measuring the headings needs to keep the north/south and east/west axes of the protractor parallel to the lines of the grid. These headings are recorded in the order of the course. The lengths of the lines are each measured and converted using the drawing's scale. The scaled length of the line is recorded with its corresponding heading. The sequential list of headings and scaled lengths provide the direction for the orienteering course.

This experience will help students to recognize that the distances in the real world and the map are relative to the map's scale. The angle measures, however, are absolute and do not change with scale. The angle measure on the map is the same as the angle measure in the real world.

Management
1. This activity is written on the assumption that students have already done *Treasure Hunt* and have orienteering skills and understanding from doing that activity.
2. This activity works well in groups of two with one student measuring the headings and lengths and the other making the scale conversion and recording the measurements.
3. This activity takes two class periods. One period is needed to draw the course and record the measurements. During the second period, students exchange their instructions and follow the course on the directions they receive.
4. Before doing this activity the teacher must find an adequate place to set up the courses. A large open field, like a playing field, works well.
5. Measure the length and width of the course and determine a scale that will allow it to be placed on a page. Suggest this scale to students or help them to determine a useful scale for themselves.

6. Prior to the activity copy the protractors onto transparency film and cut out the individual protractors.
7. Students may do this activity with or without consideration of magnetic declination. If the teacher deems that it is appropriate for the students to correct for the declination, they adjust their measured angles on the drawing by the declination for the location of the course.

Procedure

1. Discuss the *Key Questions* with students.
2. Distribute protractors, grids, and rulers to the students.
3. Make sure students know how to use the protractor and understand the scale that is to be used for the drawing.
4. Direct the students to draw the path of their treasure hunt on the grid paper using a straight edge. They need to label the *start* and *end.*
5. Beginning at the starting point, students measure the heading they would take to follow that line to the next point in the course, and record it on their record sheet. Make sure students have the north (0°) heading of their protractor facing the edge labeled *North* on their drawing, and that the north/south axis and east/west axis are parallel with the grids on the sheet.
6. Have students measure the length of the line between the two points mentioned in *Procedure 5.* Using the scale, have them convert the length of the distance that line represents on the actual course. This scaled distance is recorded with the heading on the record sheet.
7. *Procedures 5 and 6* are repeated for each line segment in the course until the end is reached. The scaled distances and the headings must be reported sequentially so the treasure hunters can find the treasures' locations.
8. Direct the students to make an answer key about their course which includes the heading the *start* is from the *end* and how far apart the *start* and *end* should be on the actual course by measuring the distance on the drawing and converting it to a ground length using the scale.
9. Have groups exchange their course directions and go out to the course site.
10. After identifying a starting point in the field, have the groups follow the set of directions they are given.
11. Have students check their accuracy of finding the treasure by using the answer key that was completed by the group that wrote the instructions.

Discussion

1. How can you determine what size to make the scale on your drawing? [Divide the width of the field by the width of the drawing space]
2. How could you make a set of directions without making a drawing? [Identify a starting point. Take the heading of where you want to go with the compass. Measure how far along this heading you go until you change headings.]
3. How could an explorer make a map of the country he was going through? [Record compass headings and distances and convert them into a scale drawing.]

Extensions

1. Have students double the distances and travel the route again. Ask them to compare this finishing point with the original set of directions. [They should be twice as far from the starting point, but in the same direction.]
2. Have students convert the distances in their instructions into the paces of one of their members (refer to *A Pace Race*). The team following the instructions gets to use the member on which the instructions were based as a "hostage" pacer.

Curriculum Correlation

Social Science

Give students a map of the western United States. Using their rulers and protractors, have them make a set of "Treasure Hunt" directions that would lead them from Independence, Missouri to Portland, Oregon. The teacher may decide if they must trace the actual Oregon Trail or follow existing roads.

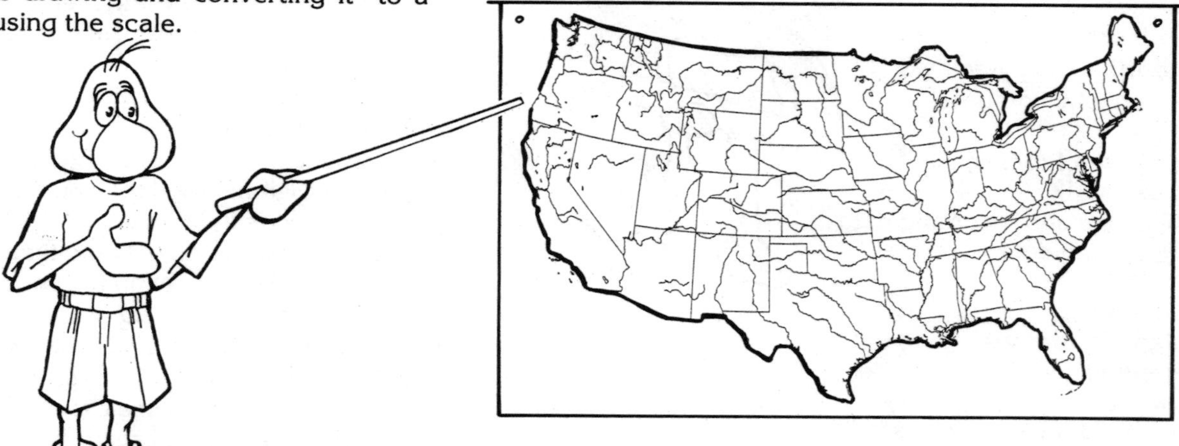

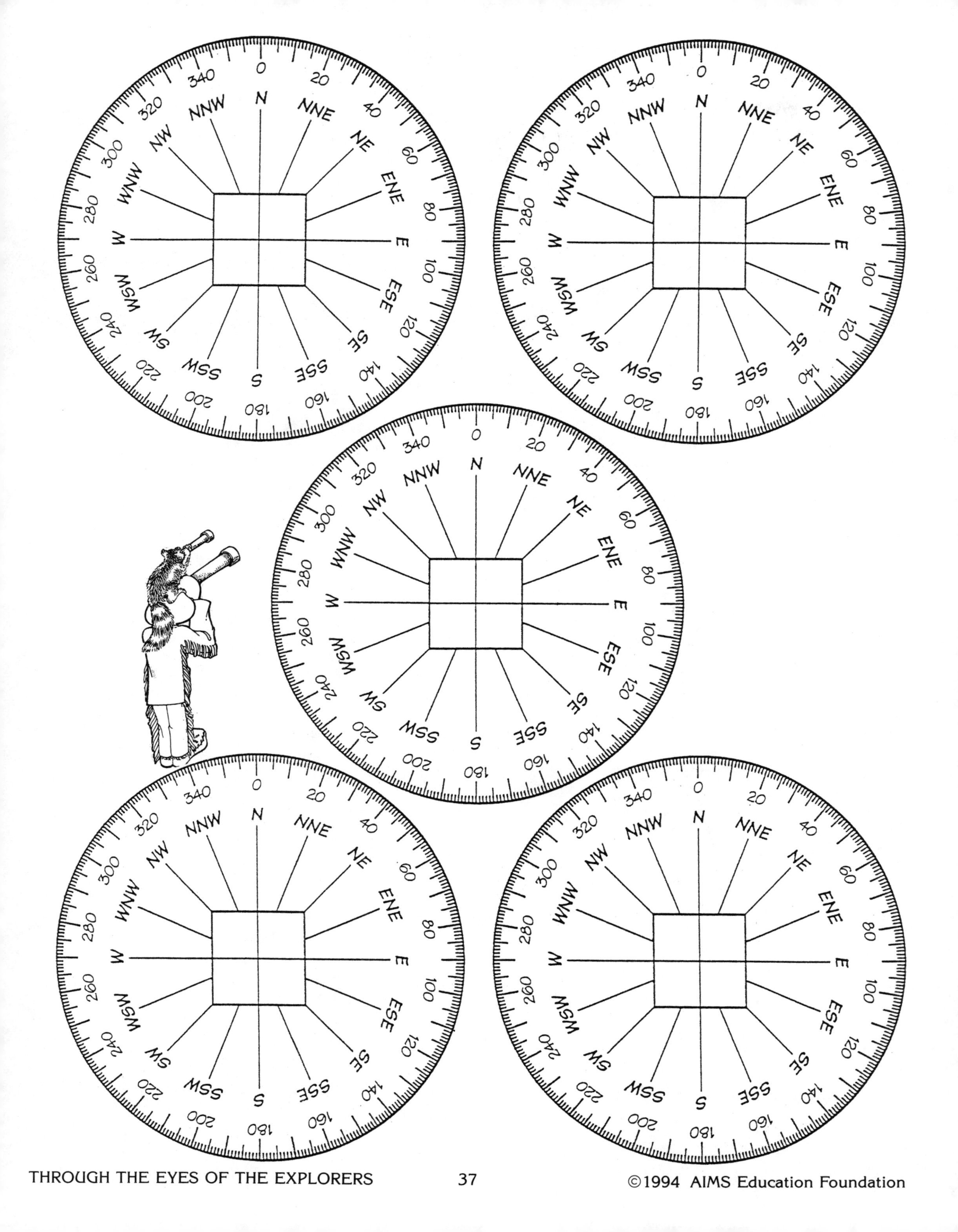
N
NNE
NE
ENE
E
ESE
SE
SSE
S
SSW
SW
WSW
W
WNW
NW
NNW
0
20
40
60
80
100
120
140
160
180
200
220
240
260
280
300
320
340

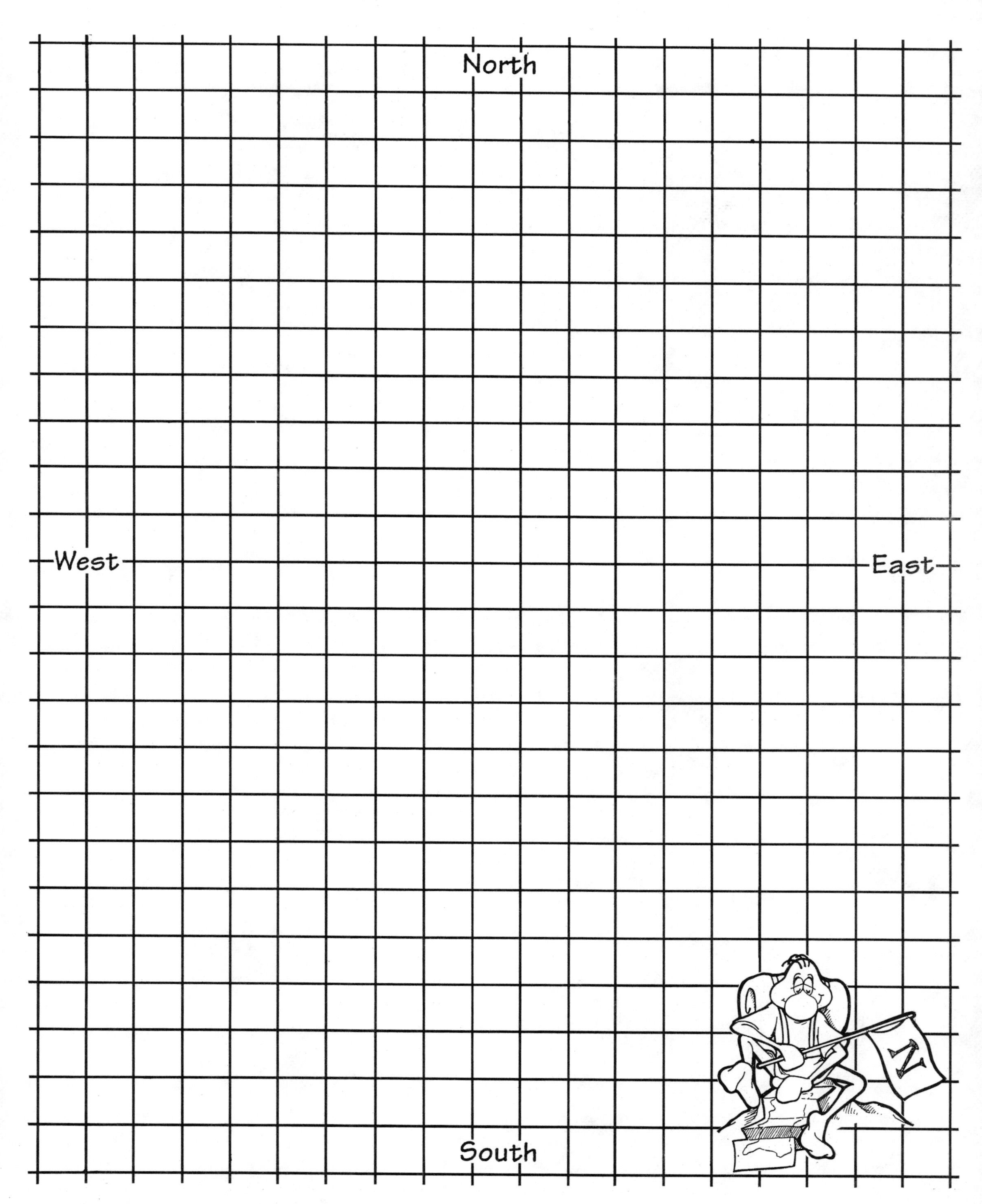
North
West
East
South
N

Instructions to find the treasure:

Start at landmark ______________

____

____

____

____

____

____

____

Here be the gold!

Instructions to find the treasure:

Start at landmark ______________

____

____

____

____

____

____

____

Here be the gold!

Instructions to find the treasure:

Start at landmark ______________

____

____

____

____

____

____

____

Here be the gold!

Instructions to find the treasure:

Start at landmark ______________

____

____

____

____

____

____

____

Here be the gold!

SUN DANCE

Topic
Apparent sun motion

Key Questions
1. How do shadows from the sun change during the day?
2. How do shadows from the sun change during the year?

Focus
Students will track shadows from the sun over a period of several months to determine patterns of apparent sun motion. The cycles they discover will provide an introduction into using the sun as a time-piece and navigational guide.

Guiding Documents
NCTM Standards
- *Explore problems and describe results using graphical, numerical, physical, algebraic, and verbal mathematical models or representations*
- *Apply mathematical thinking and modeling to solve problems that arise, such as art, music, psychology, science, and business*

Project 2061 Benchmarks
- *Because the earth turns daily on an axis that is tilted relative to the plane of the earth's yearly orbit around the sun, sunlight falls more intensely on different parts of the earth during the year. The difference in heating of the earth's surface produces the planet's seasons and weather patterns.*
- *The graphic display of numbers may help to show patterns such as trends, varying rates of change, gaps, or clusters. Such patterns sometimes can be used to make predictions about the phenomena being graphed.*
- *The motion of an object is always judged with respect to some other object or point and so the idea of absolute motion or rest is misleading.*

Math
Finding patterns
Symmetry

Science
Astronomy
  planetary motion

Social Science
Geography
Navigation

Integrated Processes
Observing
Collecting and recording data
Comparing
Predicting
Generalizing

Materials
Plain paper
Tape
Sharpened 1/4" dowel (8-12 cm long)
Board (1" x 12" x 10" minimum)
Tongue depressors
Flashlights
Optional: circular bubble level

Background Information

The shadows from the sun shining on a vertical pole fixed to flat ground show long in the morning light, grow shorter until solar noon (at which time they are the shortest) and then grow longer again until sunset. Solar noon is when the sun is highest in the sky. The shadows cast at solar noon all point north.

A record of a day's shadows on the top of a vertical pole will be symmetrical around the solar noon shadow. A shadow cast an hour after noon will be equal in length to one cast an hour before noon. If the ends of these two shadows were marked and connected by a line, one would find that the line made by the solar noon shadow will exactly bisect the line made between the two marks of the shadows cast before and after noon.

While the earth's rotation causes the daily cycle of changes in shadows, the earth's revolutions around the sun and the tilt of the earth's axis cause the length and direction of an object's shadow to change for a given hour on a yearly cycle. This causes difficulty in telling time with the sun. If one were to mark the shadow at one hour before noon today, and then come back to mark the shadow a month later at one hour before noon, the marks would form lines of different lengths and different angles from north. However, if one were to mark shadows consistently throughout the year, a consistent pattern would emerge. Such a mapping of the "sun's dance" is shown in the illustration.

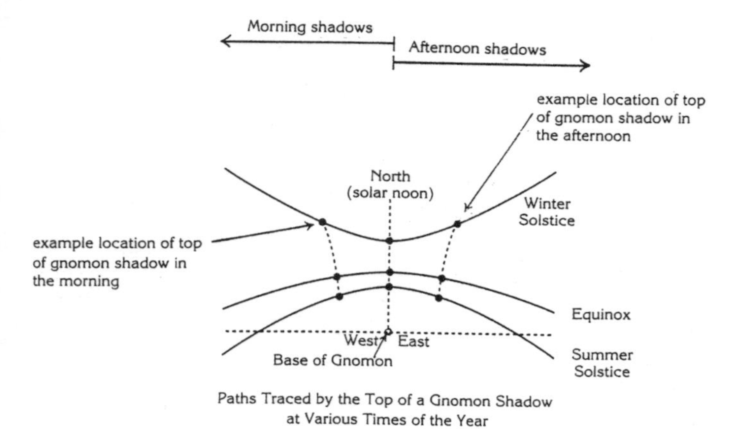

Paths Traced by the Top of a Gnomon Shadow at Various Times of the Year

In the northern hemisphere, the shortest shadow is cast on noon of the summer solstice (June 21) because the sun appears highest in the sky. During the spring and summer, the sun appears to rise and set farther north than the east-west line. As a result, the early morning and late evening shadows fall south of the east-west line. The sun is lowest in the sky and the longest noon shadow is cast on the winter solstice (December 22). The sun appears to rise and set farther south in the fall and winter months and the resulting shadow is cast to the north of the east-west line. The sun appears to rise due east and set due west only on two days of the year, the vernal equinox (March 21) and the autumnal equinox (September 22).

Management

1. This activity should be done at regular intervals (2-3 weeks) throughout the school year. Try to get a record of dates near the solstices and equinoxes. Other days of interest would be just before and after daylight savings time. Doing it on these two days will allow students to see how our clock time is relatively arbitrary.
2. This activity may be done as a small group activity or as a whole class. The use of small groups will require more boards and dowel rods. (Pencils may be substituted for dowel rods.) The whole-class method may also be done as described in *Extensions*.
3. The gnomon (dowel) must be perpendicular to the sighting surface (board). If possible, find a person (shop teacher) with a drill press and have a 1/4" hole drilled in the boards, 1" in from the center of one of the longest edges. The dowel rod can be sharpened with a pencil sharpener. If it is not possible to drill the holes, clay may be used to secure the gnomon to the paper after the paper is taped to the board; however, this is much more fragile and causes a greater error.

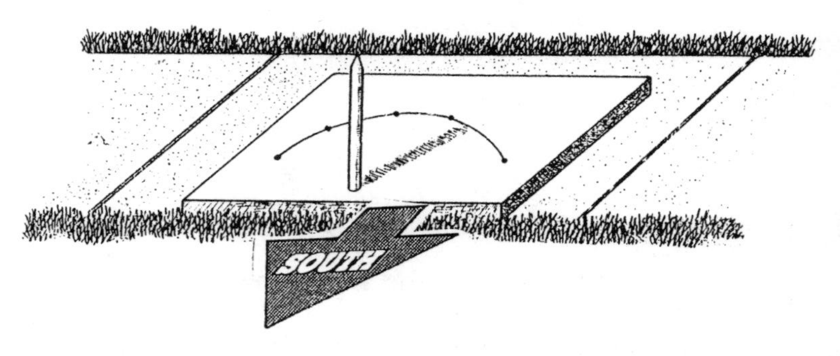

4. Find a level surface that is exposed to the sun all day throughout the year. The area will need to have a straight line or edge (i.e. the edge of a sidewalk or the side lines of a basketball court) to which the sighting board can be consistently aligned. You will need to know the general direction of north.
5. All sun sightings need to be taken on a level surface. Before the activity, check with the circular bubble level if the board is close to level when placed along the line on the surface. If it is not level, have students tape tongue depressors to the bottom edge of the board to level it.
6. Recording of shadows must be done hourly. If you do not have the same students all day, you may choose to have different groups throughout the day work on the same recording sheet or arrange for students to be dismissed from other classes throughout each day of sightings. As the year approaches the winter solstice, the shadow may become too long to do the early and late sightings each day.
7. Throughout the year, students will need to use the same length gnomon for all records, so keep each dowel or pencil with the matching record sheet and board.

Procedure

Making the Sightings

1. Discuss the *Key Questions* with students.
2. Distribute materials to groups.
3. Direct students to place the blank paper on the board by pushing it down on top of the gnomon and securing it to the board with tape.
4. Have students take the board out to the predetermined sighting area and align it with the line or edge the teacher has chosen. One edge of the board should be on this line, while the edge nearest the gnomon should be the most southern of the board's edges.

5. Direct students to make a dot on the record sheet that indicates the location of the shadow of the gnomon's point. Have them record the time by the dot and return to the classroom with the board.
6. At regular intervals (hourly) throughout the day, have students go out to the site, align their boards as before, make a dot on the record sheet where the shadow of the gnomon's point is, and record the time by the dot. After each sighting, they will need to return the board to the classroom.
7. At the end of the day, have the students make a smooth curve to connect all the day's records and record the date on the line.
8. The boards with the record sheets attached are stored carefully for another day.
9. Follow the same procedure on several different days during the year as suggested in *Management.*

Analyzing the Data

10. After at least five to six days of sightings (three months) have been completed, have students study their records to draw conclusions.
11. Have students examine the curves generated by their solar observations. Ask them to describe any trends or patterns that they notice. (They should make the following observations: all the curves are flatter around noon, the distances between dots get greater the farther from noon they are, the shadows are shortest in the middle (noon), the curves are symmetrical around the middle, the curves are flattest at the time of the equinox and get more convex and concave as the date from the equinox increases.)
12. Have students connect all the 10:00 readings, all the 12:00 readings, and all the 2:00 readings. Have them describe what patterns they see and what inconsistencies there are in the patterns. (Students should notice that the connected readings form smooth curves. These curves bend away from the center line as you approach the winter solstice. The curves all shift one hour to the right when the clocks change to standard time, and shift to the left when the clocks change to daylight savings time.)
13. Distribute a flashlight to each group and darken the room.
14. Have students use the flashlight to make the gnomon's point cast a shadow to specific sighting marks on their record sheets. Many students will need to explore and practice this skill. Encourage students to recognize that the length of the shadow is altered by changing the elevation of the flashlight, not by changing the flashlight's distance from the gnomon.
15. Have a student in each group adjust the elevation of the flashlight so the shadow point moves to all the sightings of a given hour. Tell the students that the flashlight is reproducing the position of the sun as these sightings were made. As the other students watch, have them describe the motion of the flashlight as if it were the sun. (Students should recognize that the sun's angle gets lower as the winter solstice is approached and higher as the summer solstice is approached.) Have students repeat this procedure for other hours to confirm their findings.
16. Have students take turns being the "sun" with the flashlight. Have them try to follow the sun's apparent motion for each day by making the shadow of the gnomon's point trace the track for each day.
17. Have students predict where the track would be if taken several weeks later.

Discussion

1. When are the shadows shortest? [around noon standard time or 1:00 p.m. daylight savings time]
2. Why is the shadow shortest at noon? [sun appears highest in the sky]
3. If you went out today, how could you determine when it was solar noon without a watch? [shortest shadow]
4. When are the noon shadows shortest? [summer solstice]
5. When are the noon shadows longest? [winter solstice]
6. Why does the length of the noon shadow change? [elevation of sun changes seasonally]
7. What patterns do you see in the curves you recorded?
8. If you were shown a curve formed by recording the movement of a shadow, how could you determine what season it was? [Match the curve to those you recorded. Each curve has two possible matches, one on each side of an equinox (fall and winter match, and spring and summer match).]
9. Why do these patterns show up? [consistent movement of the earth, sun's apparent motion]
10. Describe how the apparent movement of the sun has changed over the period of the sightings.

Extension

You may choose to do this activity as a whole class activity. Find a vertical pole that is secured in the ground with a level surface of concrete or asphalt to the north of pole. A fence post along a sidewalk, or a basketball standard pole on a playground would work well. Follow the general instructions in the Procedure but permanently mark the positions of the shadow with paint. Since students cannot use the flashlight with this method, they may choose to position their heads at different sighting points. As they look towards the top of the pole, they will see where the sun would be at that time and can compare the differences at other points.

Topic
Magnetic declination

Focus
By using a sun sighting to determine geographic north, students will compare geographic north to a compass' magnetic north to determine the magnetic declination at their position.

Key Questions
1. When does the sun's shadow point north?
2. How could you find true (geographic) north so you can check the accuracy of your compass?

Guiding Documents
NCTM Standards
- *Explore problems and describe results using graphical, numerical, physical, algebraic, and verbal mathematical models or representations*
- *Apply mathematical thinking and modeling to solve problems that arise, such as art, music, psychology, science, and business*

Project 2061 Benchmarks
- *Because the earth turns daily on an axis that is tilted relative to the plane of the earth's yearly orbit around the sun, sunlight falls more intensely on different parts of the earth during the year. The difference in heating of the earth's surface produces the planet's seasons and weather patterns.*
- *The graphic display of numbers may help to show patterns such as trends, varying rates of change, gaps, or clusters. Such patterns sometimes can be used to make predictions about the phenomena being graphed.*
- *The motion of an object is always judged with respect to some other object or point and so the idea of absolute motion or rest is misleading.*

Math
Measurement
  angle
Geometric constructions

Science
Earth science
  astronomy

Social Science
Geography
  navigation

Integrated Processes
Observing
Collecting and recording data
Interpreting data
Generalizing
Applying

Materials
Record sheet
Tape
Sharpened 1/4" dowel (8-12 cm long)
Magnetic compass
Board (1" x 12" x 10" minimum)
Circular bubble level
Tongue depressors
Chalk
Straightedge
Drawing compass

Background Information
The shadows from the sun show long in the morning, grow shorter until solar noon and then grow longer again until sunset. The shadows are shortest at solar noon when the sun is highest in the sky. The shadows cast at solar noon all point north.

A record of a day's shadow will be symmetrical around the solar noon shadow. A shadow cast an hour after noon will be equal in length to one cast an hour before noon. If one were to mark the ends of these two shadows during the day, and then connect the two marks, one would find that the line made by the noon shadow will exactly bisect the line made between the two marks of the shadows cast before and after noon.

Such a method has been used historically to establish geographic north. A stick is placed in the ground in a vertical position. At sometime before noon, the end of the stick's shadow is marked on the ground. A circle is then drawn on the ground with the stick as the center and the end of the shadow on the circle. The shadow grows shorter until noon, when it begins to lengthen. The end of the shadow touches the circle again when it is exactly the same time differnece after noon as it was before noon when the initial mark was made. By bisecting the line connecting the two marked points on the circle and connecting this bisection point with the center of the circle, one has established a true north and south line.

Management

1. Prior to this activity, students should have done *Sun Dance* so they have an understanding of the earth's motion and its effect on shadow lengths. This activity may be done in conjunction with *Where on Earth*? The same record sheet may be used for each activity if the near noon sightings are done as instructed in *Where on Earth*?
2. This activity may be done as a small group activity or as a whole class. Small groups will require more boards and dowel rods. (Pencils may be substituted for dowel rods.) The whole class method may also be done as described in E*xtensions*.
3. The gnomon (dowel) must be perpendicular to the sighting board. If possible, drill a 1/4" hole in each board, 1-inch in from the center of one of the longest edges. The dowel rod can be sharpened in a pencil sharpener before the blunt end is pushed into the drilled hole. If a hole cannot be drilled, clay may be used to secure the gnomon to the record sheet after it is taped onto the board; however, this is much more fragile and causes greater error.
4. Find a surfaced area that is generally level and has all day exposure to the sun. Be sure of the general compass heading at this site. The procedure assumes that the area is not secured throughout the day and that the sighting boards may be disturbed if left outside. If this is not the case, the area does not need to be surfaced, and the sighting boards may be left at their positions all day.
5. It is critical when making measurements from sun sightings that they be taken on a level surface. Students can tape tongue depressors to the bottom edge of the sighting board to level it. This is crucial, make sure they do it carefully!
6. The shadow observations and recordings must be done hourly. If you do not have the same students all day, you may choose to have different groups work on the same recording sheet and board, or arrange for students to be dismissed from other classes throughout each day of sightings.

Procedure

1. Discuss Key Q*uestions* with students.
2. Distribute materials to groups.
3. Have students tape the record sheet to the sighting board, pushing it down on top of the gnomon at the labeled position.
4. Direct the students to go out to the sighting area and choose a place to put their sighting boards. They need to align the boards so the broken line runs generally east and west, and the semicircles are north of the broken line.
5. Referring to a circular bubble level placed on top of the sighting board, have students tape tongue depressors to the bottom of the board until the board is level.
6. When the students are sure of the position and alignment of their board and have leveled it, have them use chalk to draw an outline of it on the pavement with chalk. Direct them to label the outline so they can return to the same spot all day.
7. Have students place a dot on the record sheet where the shadow of the gnomon's point is, record the time by the dot, and return to the classroom with the sighting board.
8. At regular intervals (hourly) throughout the day, have students take their sighting board out to the area and place it in its original position as marked by the chalk outline. Each time they will need to make a dot on the record sheet where the shadow of the gnomon's point is and record the time by the dot.
9. At the end of the day, or at the beginning of the next day, have the students follow the instructions on the student page to draw the line of geographic north, (the line of the shortest shadow of the day, the shadow of solar noon).
10. Invite the students to return to the original position of their sightings and align the board with the chalk outline.
11. Have them place a compass on the record sheet so a directional arrow on the base is on and parallel with the line of geographic north.
12. Direct students to rotate the compass bezel (the rotating dial of the compass) until the North-South arrow is pointing north, and is directly under the needle.
13. Have them determine the difference (declination) from the geographic north of their sighting, to the magnetic north of the compass by referring to the degree markings on the bezel.

Discussion

1. What patterns do you see in the curve you recorded? [symmetry, looks the same on both sides]
2. Why does this pattern show up? [consistent movement of earth, sun's apparent motion]
3. Where are the shadows the shortest? [towards the middle]
4. About what time of day is it when the shadows are shortest? [noon]
5. Which way do these short shadows point? [generally north]
6. What did the line you drew from the gnomon do to the shadows' curve? [cut it in half, made a line of

symmetry]

7. If the gnomon cast a shadow along this line of symmetry, what would be true? [It would be the shortest shadow.]
8. At what time of day are shadows made by the sun shortest? [at solar noon]
9. What direction should this shortest noon shadow be pointing? [straight north, geographic north]
10. What did the line segment drawn between the gnomon and the arc intersection do to the line segment between the intersections of semicircle and the shadow's curve? [cut it in half at right angles, perpendicular bisector]
11. How could you determine what is due east? [Use the line-segment between the intersections of the semicircle and the shadow's curve.]
12. If you went out tomorrow, how could you determine when it was noon without a watch? [shortest shadow, shadow lines up on north line]
13. How does the north line made by the gnomon's shadow differ from the north of the compass? (Answers will differ by geographic location. For the amount of magnetic declination at a location, refer to the *Background Information* in the activity *Treasure Hunt.*)
14. How would you use this knowledge of the difference in geographic north and magnetic north if you wanted to go "straight north" and had a compass? [Adjust for magnetic declination.]

Extensions

1. You may choose to do this activity as a whole class activity. Find a vertical pole that is secured in the ground with a level surface of concrete or asphalt to the north of pole. A fence post along a sidewalk, or a basketball standard pole on a playground would work well. Follow the general instructions in the procedure, but mark the positions of the shadow with chalk. To find north, you may use string and chalk to draw a circle with the pole as the center. Continue to use the string and chalk as a drawing compass to do the geometric construction to find geographic north.
2. Have students research and find out the exact magnetic declination for their location. This information can be found on a recent geological survey map, aeronautical map, or by calling a U.S. Dept. of Transportation Flight Service Station.

Special Instructions

1. *Draw a smooth curve to connect the dots that mark the ends of the shadows on your record sheet.*

2. *Find a semicircle on the record sheet that your curve intersects twice. Place X's at the intersections of the curve and semicircles.*

3. *Using a straightedge, draw a line segment between the two intersections you marked with X's.*

4. *Spread a drawing compass so the distance between the stylus (pointed end) and the pencil point is more than half the length of the line segment.*

5. *Place the stylus point on one end of the line segment (circle-curve intersection) and draw an arc of a circle on the side of the line segment that is opposite the gnomon.*

6. *Without adjusting the drawing compass, place the stylus on the other end of the line segment and draw an arc on the same side of the line segment as the first arc.*

7. *With a straightedge, draw a line segment from the intersection of the two arcs to the center of the gnomon.*

8. *When the sighting surface is aligned in its original outside position, this final line segment will be aligned between geographic (true) north and south.*

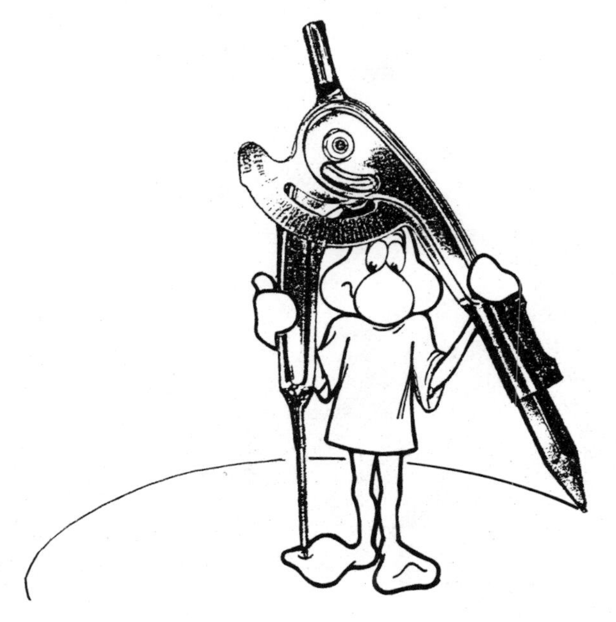

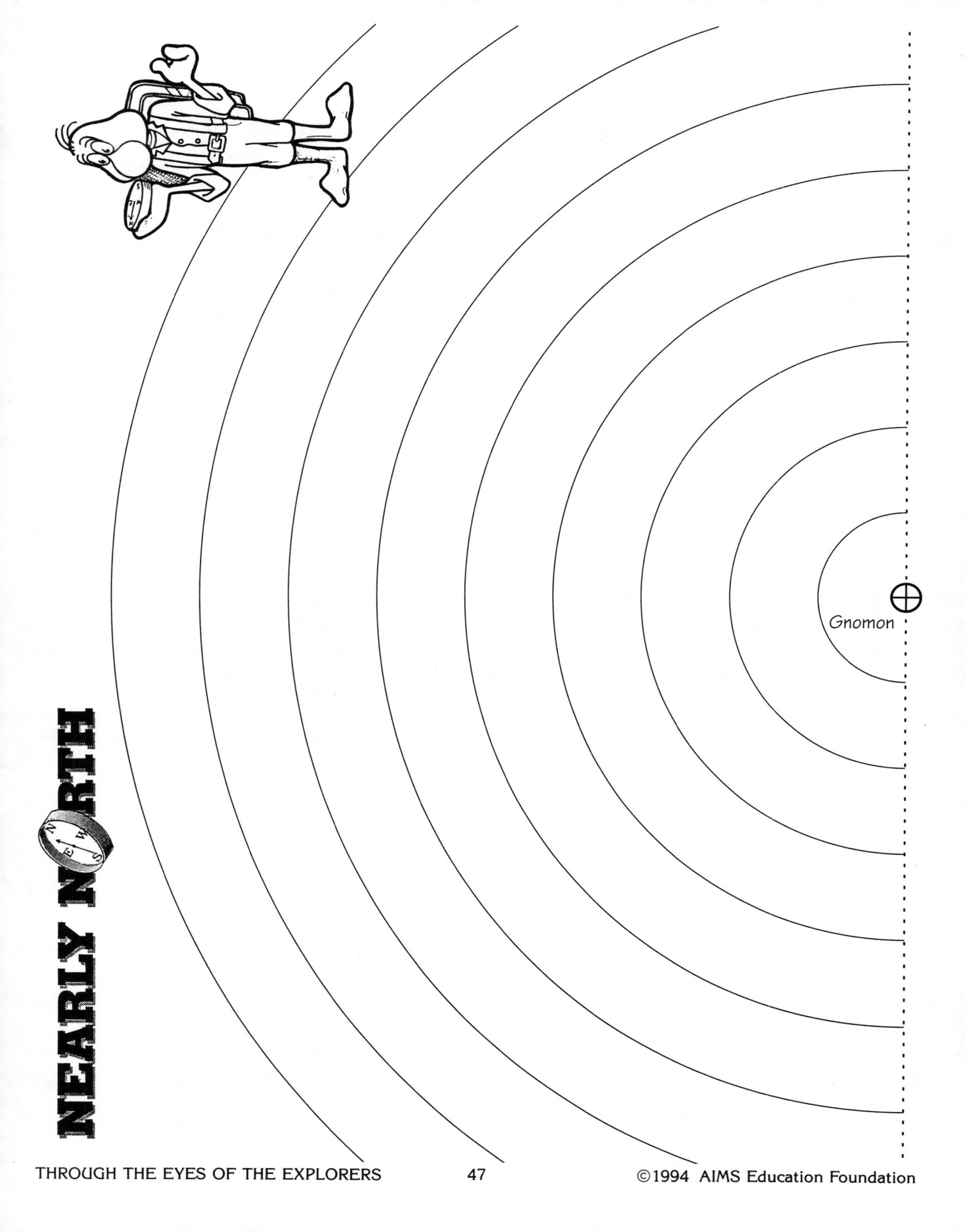
NEARLY NORTH
Gnomon

WHERE ON EARTH?

Topic
Determining latitude and longitude

Key Question
How were the explorers able to use the stars to know where they were on the earth?

Focus
Students will use a shadow made by the sun at solar noon to determine their latitude and longitude.

Guiding Documents
NCTM Standards

- *Develop an appreciation of geometry as a means of describing the physical world*
- *Value the role of mathematics in our culture and society*
- *Develop and apply a variety of strategies to solve problems, with emphasis on multi-step and non-routine problems.*

Project 2061 Benchmarks

- *Mathematics is helpful in almost every kind of human endeavor–from laying bricks to prescribing medicine or drawing a face. In particular, mathematics has contributed to progress in science and technology for thousands of years and still continues to do so.*
- *It takes two numbers to locate a point on a map or any other flat surface. The numbers may be two perpendicular distances from a point, or an angle and a distance from a point.*
- *The motion of an object is always judged with respect to some other object or point and so the idea of absolute motion or rest is misleading.*

Math
Measurement
  linear
  time
  angle
Reading graphs
Using formulas
Integers

Science
Astronomy
Planetary motion

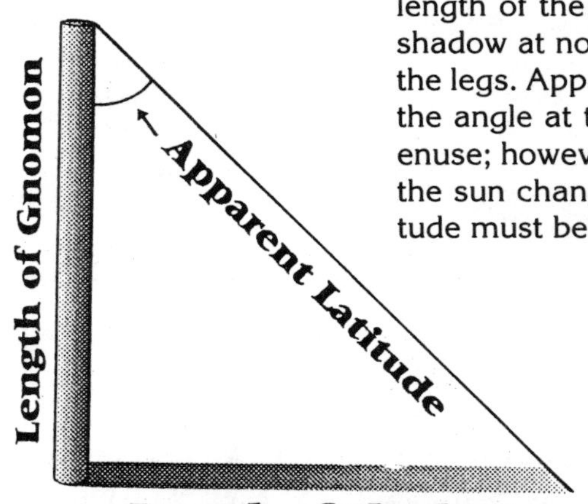

Social Science
Geography
Navigation
Latitude and longitude
Time zones

Integrated Processes
Observing
Collecting and recording data
Applying

Materials
Tape
Sharpened 1/4" dowel (8-12 cm long)
Board (1" x 12" x 12" minimum)
Circular bubble level
Tongue depressors
Magnetic compass
Watch with second hand
Blank sheet of paper
Straightedge
Student pages
Map of area with longitude and latitude

Background Information
Latitude tells the number of degrees north or south of the equator a position is. Longitude tells how many degrees east or west of the Prime Meridian a position is. By knowing the exact time and angle of the sun at solar noon, a position's latitude and longitude may be determined.

To determine solar noon, a *gnomon* (vertical stick for casting a shadow) is placed in the ground. When the gnomon's shadow is shortest and is cast due north, it is solar noon. At solar noon, the shadow's end is marked and the exact time is noted.

To find latitude, a triangle is drawn with one leg the length of the gnomon, the other leg the length of the shadow at noon, and the hypotenuse joins the ends of the legs. Apparent latitude is determined by measuring the angle at the vertex of the gnomon and the hypotenuse; however, due to the tilt of the earth, the angle of the sun changes on a yearly cycle. The apparent latitude must be corrected to account for this change.

Longitude can be determined because of its direct correlation to time. The earth turns 360° in 24 hours, or 15° an hour, or .25° a minute. This means that when it is solar noon at 90° west longitude, it is 6:00 p.m. at 0° longitude. If the time at the Prime Meridian (Greenwich Mean Time) is known when it is solar noon at a different longitude, then the longitude can be calculated by multiplying the hours past noon Greenwich Mean Time (GMT) by 15° per hour.

The standard time zones are each 15° wide (allowing for exceptions of political boundaries) or 1 hour of sun time wide. The Pacific Time Zone is centered on 120° west longitude which is eight hours sun time west of Greenwich. When solar noon comes at 12:00 Pacific Standard Time, the longitude reading would be 120° west.

If solar noon occurred before or after 12:00 clock time, the longitude readings would be slightly east or west of the 120° west longitude. To calculate the position, noon standard time is subtracted from local solar noon. This difference is multiplied by .25° per minute. The answer to the calculation is added to the zone's degrees west of Greenwich. (With a negative number, this becomes a subtraction problem.)

The sun, however, is not a perfect time piece. The earth's elliptical orbit makes sun time accurate only 4 times a year. To be accurate in determining longitude from sun time, seasonal adjustments must be made.

People have been able to determine their positions north or south for a long time. Oral traditions support the fact that prehistoric humans were aware of their position by watching the stars, including the sun. The ability of determining east and west positions is a relatively recent development due to the need to keep time accurately.

During the 17th century, with the use of telescopes, it was discovered that time could be kept by watching the moons of Jupiter. This was of little help to ship navigators being tossed at sea. In 1707 when the British fleet lost four ships and 2,000 men on the Scilly Islands, it became obvious that there was a need for accurate, portable clocks to determine longitude. In 1714 the British government encouraged the invention of such a clock by promising an award of 20,000 £ (more than $20,000). John Harrison, a carpenter and tinkerer, spent over 30 years of his life developing a clock. In 1761, he finally got one to work that kept time to within .04 seconds a day.

With the development of the accurate clock and British control of the seas and commerce, the British way of mapping became standard throughout the world. They decided that the Prime Meridian should run through Britain. The city of Greenwich, a suburb of London, was chosen because it was the site of the Royal Observatory. The English placed themselves in the center of map, so to speak.

Time zones are another topic of historical interest. Until 1883 each town had its own time set to solar noon at its longitude. With literally thousands of times, it was chaotic for the train companies trying to make schedules. On November 18, 1883, the clocks around the country were set to standard time zones. For the first time in history, clocks determined the time, not the sun. It took until 1917 for the United States government to make Standard Time the law.

Clocks again changed life styles with the introduction of Daylight Savings Time. It was first tried in Germany during World War I as an attempt to save energy. The Time Act of 1917 introduced it to this country. Daylight Savings Time lasted only a year and a half. People did not like it. It was introduced again during World War II. Today it continues, but is a decision made by more local jurisdictions.

Management

1. You should plan to do this activity after the completion of *Sun Dance* and *Nearly North* so students will be familiar with the procedures and understand why they are doing them.
2. From the data gathered in *Sun Dance* or *Nearly North*, or from your own observations, determine when solar noon occurs.
3. The sun waits for no one. This activity must be done at solar noon. If your class is not in session, you will need to make arrangements to do this activity through release time or as a special session if noon falls during lunch break.
4. Record data for about a half an hour, 15 minutes on either side of your predicted solar noon. If you are not sure, run your records for an hour centered around 12:00 noon Standard Time.
5. You will need to have accurate time on your watch. Use the telephone to establish the correct time. (The phone number for time is listed in the white pages under *Time.*) Adjust your watch accordingly and make sure to adjust the time to Standard Time if your area is on Daylight Savings Time. To provide students with an accurate time during the activity, you will need to call out the time (in minutes) for the students.
6. This activity is designed to be done in small groups. If you want to do it as whole class activity, refer to *Extensions* in *Nearly North* and modify to fit.
7. Find a level area that is exposed to the sun during noon time. Refer to *Management* in *Nearly North* on how to construct the sighting boards.
8. A chart is included on the bottom of the student sheet on longitude for correction due to the earth's

elliptical orbit. You may choose to delete it if you do not feel it is appropriate for your students. If you have your students use it, have them add the correction time from the difference in zone mean time before they calculate the degrees.

9. The procedure has students orient the sighting board using a compass. If students do this activity immediately after doing *Nearly North*, they may leave their record sheet from *Nearly North* attached to the sighting board and return to their labeled position in the sighting area. Aligning the board with the labeled position will align the north-south line with geographic north. Another alternative is to do this activity on the day *Nearly North* is done. Have students come out near solar noon and put the points down as described in *Procedures 6-8* below. They can complete the rest of the *Procedure* at a later time.

Procedure

1. Discuss the *Key Question* with students.
2. Distribute materials to groups and discuss the need to be on Standard Time if you are on Daylight Savings Time.
3. Have students make a group record sheet by drawing a straight line across the width of their blank paper to serve as the north-south line.
4. Direct students to tape their record sheet to the sighting board, pushing it down on the gnomon so the gnomon goes through the north-south line.
5. Have them take the sighting board out to the predetermined sighting area, setting it on the ground so the line is aligned north and south with the gnomon on the southern edge of the board, and adjusting the board until it is level. Use a magnetic compass to accurately align the north-south line with geographic north. Students must adjust for magnetic deviation for your area (refer to *Treasure Hunt*). Use a circular bubble level placed on the board to determine if the board is level. Tongue depressors can be slid under the board to help level it. Have the students recheck the board's alignment after leveling the board.
6. Direct the students to place a dot on the record sheet where the shadow of the gnomon's point is and record the time by the dot.
7. For each minute interval, announce the time and have students make a dot on the record sheet where the shadow of the gnomon's point is, and record the time by the dot.
8. When the gnomon's shadow has gone several minutes to the east of the north-south line, students may stop recording.
9. Have students determine the point that is closest to being directly on the north-south line. Have them note the time this point was recorded. This is solar noon.
10. Direct the students to measure (to the nearest millimeter) and record the length of the shadow at solar noon (the dot closest to the north-south line) and the height of the gnomon.
11. Using the data from their sighting and following the instructions on the student pages, have students determine their longitude and latitude position.
12. Have students refer to a map to determine their longitude and latitude and compare their prediction with the map's information.

Discussion

1. What did your group calculate our latitude and longitude to be?
2. Looking on the map, what is our latitude and longitude?
3. How might you get a more accurate reading if you did this again? [Averaging all the groups' longitudes and latitudes will give more accurate readings.]

Extension

1. Try the activity at different times of the year and see how your apparent position changed. You could use your data from *Sun Dance*.
2. Research how sundials are made and use your calculations to make a sundial appropriate for your latitude.
3. Research navigational techniques for determining latitude and longitude. How are they similar and different from those in this activity?

WHERE ON EARTH?

LATITUDE

1. Record on the vertical ruler below the length of your gnomon.
2. Record on the horizontal ruler below the length of the gnomon's shadow at solar noon (the shortest length or when the shadow points straight north).
3. Draw a line connecting the endpoints of the two lines you drew.
4. Measure the angle in the top corner of the the triangle you have drawn—the angle at the top of the line that represents your gnomon. Record the measure.

The measure of this angle tells you how many degrees you are from directly under the sun. It tells you how many degrees north or south you are of the sun's zenith (straight overhead). The degrees north or south you are from the equator is called your latitude.

The earth is tilted on its axis. As a result, the sun's zenith moves back and forth across the tropics on a yearly cycle. In the winter the sun appears lower in the sky because the sun's zenith is farther south, and in the summer the sun appears higher in the sky as its zenith moves north. The movement of the sun's zenith will change the length of the noon shadow throughout the year. This will change the angle you measured in step 4, apparently changing your latitude throughout the year. Use the graph to determine how much correction must be made to the latitude you measured.

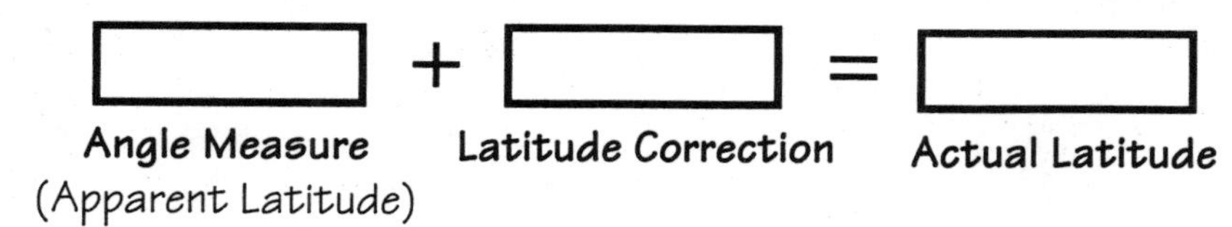

Latitude Correction Graph

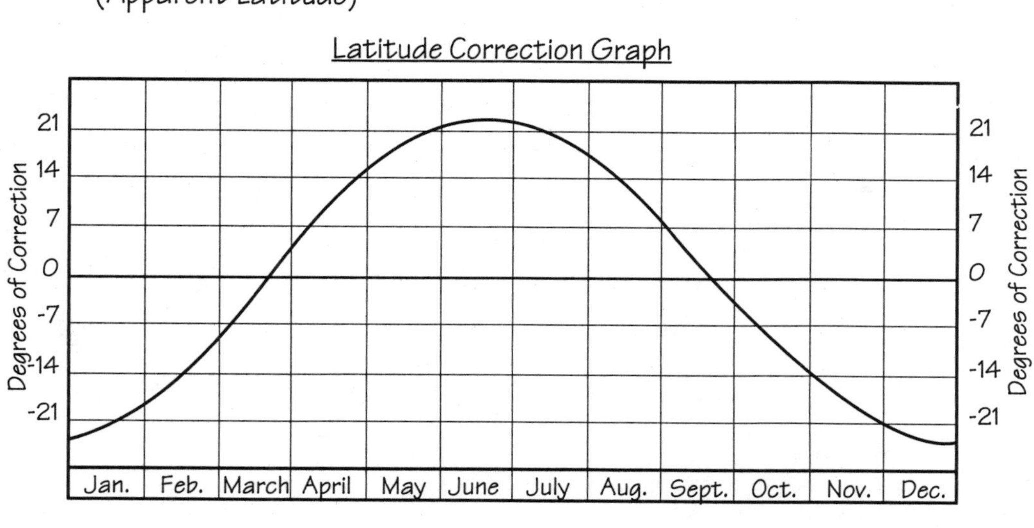

Gnomon's Height (cm)

0 1 2 3 4 5 6 7 8 9 10 11 12 13 14 15

Shadow's Length (cm)

0 1 2 3 4 5 6 7 8 9 10 11 12 13 14 15 16 17

LONGITUDE

There are 360 degrees of longitude around the earth. The earth rotates once in 24 hours. How many degrees does the earth rotate each hour?

If the earth rotates 15 degrees an hour and there are 60 minutes in an hour, how many degrees does the earth rotate in a minute?

Lines of longitude are numbered from the **Prime Meridian** which runs through Greenwich, England (a suburb of London). It is noon in Greenwich, England, when a shadow made by the sun is cast directly north. Since you know the earth's degrees of rotation per hour, you should be able to complete the chart to the right to determine how many hours later noon is in each time zone in North America than noon is in Greenwich, England.

Time Zone	Degrees West of Greenwich	Hours West of Greenwich
Atlantic	60	
Eastern	75	
Central	90	
Mountain	105	
Pacific	120	
Yukon	135	
Alaska/Hawaii	150	

Degrees of longitude from Greenwich can be turned into solar time, and solar time can be turned into degrees.

How many hours later is noon for you than Greenwich, England?

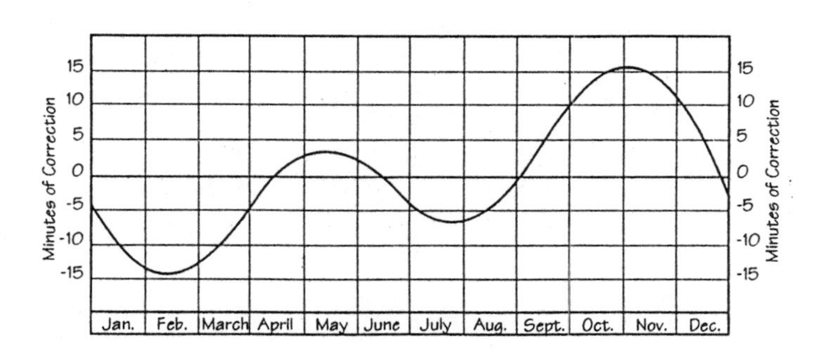

☐	Solar Noon (Standard Time)
− 12:00	Zone Noon (Standard Time)
☐	Difference from Zone Noon (in minutes)

☐	x	.25	=	☐
Difference from Zone Noon		Degrees per Minute		Degrees of Difference
☐	+	☐	=	☐
Zone's Degrees West of Greenwich		Degrees of Difference		Degrees of Longitude

About how many degrees west of Greenwich do you live?

A time zone is a band of longitude 15 degrees wide. If you live in the center of the time zone, your solar noon is the same as noon on your watch. You probably found that your solar noon was a little before or after noon on your watch. Use your solar noon time to make the calulations to the left and determine what your exact longitude is.

SIZING UP SHADOWS

Topic
Indirect measurement

Key Question
How can you determine the height of an object from its shadow?

Focus
Students will measure shadow lengths cast by dowels of various heights to establish the understanding that the length of a shadow and the height of the object have the same proportion for all objects at any given time.

Guiding Documents
NCTM Standards

- *Develop formulas and procedures for determining measures to solve problems*
- *Understand and apply ratios, proportions, and percents in a wide variety of situations*
- *Analyze functional relationships to explain how a change in one quantity results in another*
- *Use patterns and functions to represent and solve problems*

Project 2061 Benchmarks

- *Organize information in simple tables and graphs and identify relationships they reveal*
- *The graphic display of numbers may help to show patterns such as trends, varying rates of change, gaps, or clusters. Such patterns sometimes can be used to make predictions about the phenomena being graphed.*

Math
Proportional reasoning
Patterns and functions
Measuring
  length
Graphing
Averaging

Science
Terrestrial motions
  light and shadows

Integrated Processes
Observing
Collecting and recording data
Interpreting data
Drawing conclusions
Applying and generalizing

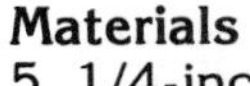

Materials
5 1/4-inch dowels
Saw or sharp knife
Masking tape
String
1" grid easel paper
Tape measures
Felt tip markers
Scissors

Background Information
At any given time of the day, all the ratios of the heights of objects to the length of their shadows will be the same. This happens because sunlight travels in parallel lines. For every object, the ground, the rays of light, and the line perpendicular to the ground through the highest point of that object form a triangle. These triangles are similar for all objects because all their corresponding sides are parallel.

The proportional aspect of these similar triangles makes it possible to determine the height of a tall object. A smaller object and its shadow are used to determine the proportion of object height to shadow length. This ratio is then applied to the tall object's shadow to determine its height.

Management

1. This activity must be done on a sunny day on a level surface which has unobstructed sunlight during the activity period. The area should be in the proximity of a tall object such as a flag pole, light pole, telephone pole, basketball goal, etc. which will cast a shadow onto level ground.
2. The data is gathered by eight or nine small groups. The rest of the activity is done as a class with the groups sharing their data.
3. Prior to the activity, prepare the five dowels as illustrated. Eight rods of differing heights are needed (see *Figure 1*). Each rod needs an additional length of about 10 or 11 centimeters so they can be imbedded in the ground.

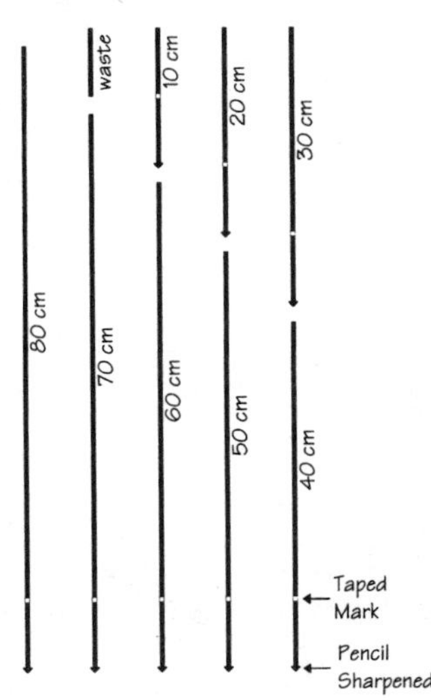

4. From each end of the dowel, measure in one of the given heights for that dowel. Put several wraps of masking tape at these points. Allow for the inground portion and cut the dowels. (A pencil sharpener can be used to make them easier to push into the ground.)
5. Prepare easel paper for class graph by taping it so its longest dimension is horizontal. With a marker draw in the horizontal axis (x-axis) along the bottom and label it *Shadow Length*. Draw in the vertical axis (y-axis) along the left edge. Label it *Height of Stick*. Depending on the length of the shadows produced, more paper may need to be added to extend the graph horizontally.
6. Steps 20-24 are optional. They are included to help reinforce an understanding of similar triangles, to help make a connection of proportionality with the students' concepts of simultaneous additions, and to provide alternative solutions to the problem.

Procedure

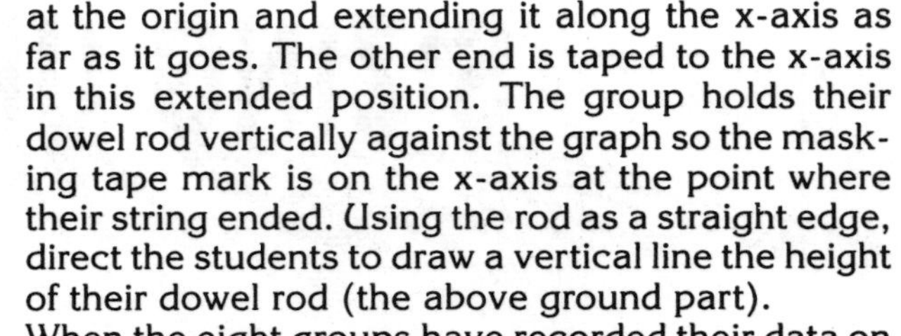

1. As a class discuss the *Key Question* along with questions such as:
   How do shadows change during the day?
   How do shadows change with the seasons?
   How will the shadows from these sticks be different from each other?...similar to each other?
2. Divide the class into eight or nine groups. Distribute a dowel rod, scissors, and a length of string long enough to measure the shadow of the group's rod. The ninth group will not get a rod but will measure the shadow of a designated tall object in the area. They will need the appropriate amount of string for that object's shadow.
3. Take the students to the selected area.
4. Have the students push the sharpened end of their dowels into the ground until the masking tape mark is even with the earth's surface. Have them sight all around the dowels to check that they are perpendicular to the ground and make adjustments if they aren't.
5. At a given signal from the teacher, one student from each group should place a finger at the end of the rod's shadow and keep it there until the other members lay out their string along the shadow line and cut it to that length. (If there is not a ninth group, the teacher needs to mark the end of the tall object's shadow and get assistance in cutting the string.)
6. Return to the classroom with strings and rods to share data.
7. Starting with the group with the longest rod, have each group come up one at a time and place their data on the graph by taping one end of their string at the origin and extending it along the x-axis as far as it goes. The other end is taped to the x-axis in this extended position. The group holds their dowel rod vertically against the graph so the masking tape mark is on the x-axis at the point where their string ended. Using the rod as a straight edge, direct the students to draw a vertical line the height of their dowel rod (the above ground part).
8. When the eight groups have recorded their data on the graph, ask the students what pattern they see. They will recognize that the dowels are evenly spaced and that the top of the dowels form a line.
9. Draw in this line, extrapolate it to the origin. Ask the students what this line represents? [The line of the sunlight; *The slope of this line also represents the ratio of the height of the object to its shadow's length.*] Ask students what shape they see with every rod, their string, and the line that was drawn. [triangle] Ask them how the eight triangles are alike. [same slope, similar triangles]
10. Have a volunteer from each group come up to the graph with their dowel and compare the stick's height to the shadow's length. Encourage students to physically measure their dowel against their string. Have them determine how many shadows long the rod's length is. Fractions will occur if the following ratio is made:

$$\frac{\text{1 shadow length}}{\text{rod's length, in numbers of shadow lengths}}$$

(If the shadow is shorter than the height, they will get a fraction less than 1. If the shadow is longer, they will get a number greater than 1.)

11. Ask students to explain where they think the tall object would be on the graph. Ask students how they could use the string from the shadow of the tall object to determine its height.
12. Have a volunteer from each group, one at a time starting with the group with the longest rod, come up to measure their *Shadow Length* and *Height of Stick* to the nearest centimeter. Have them report their measurement to the whole class so students can record the data on their charts.
13. Have students make a coordinate graph of the data on their chart.
14. Have students compare the height to the shadow by completing the ratio column. The students should notice that this ratio is very similar for each of the dowels. Have the students calculate an average ratio. Remind students of their comparison of the length of the rod to the length of the string in *Procedure 10*. Their calculations provide the same information, but more accurately.
15. Have the students state in their own words what this ratio is telling them about the height of the stick and its shadow.
16. Ask how this ratio describes the line in the graph. [Every time the line goes over one space to the right, it goes up the number of spaces of the ratio. The slope of the line is the ratio.]

17. Ask students how they would use the graph to predict the height of an object that casts a 65 cm shadow...100 cm shadow...the length of the tall object's shadow.
18. Measure the length of string representing the tall object's shadow and record on the bottom row of the chart.
19. Have students calculate the height of the tall object.

Optional (20-24):

20. Have students discuss what patterns they see in the chart. (Students should recognize that height goes up in 10 centimeter increments.) Have students find and record the difference between each of the adjacent *Shadow* lengths. At this point, students will recognize that the shadows also consistently change.
21. Ask the students to determine how these two columns of consistent differences show up on the graph. [Evenly spaced rods, and the rods get longer by the same amount each time]
22. Have students discuss why adding consistently different amounts to the height and the shadow keep the triangles congruent. [Example of responses: The measurement that started bigger needs to add a bigger piece to keep its *proportional lead.* If both sides were to grow twice as big, the bigger one would have added more length to its side]
23. Have students state what these two columns of differences tell about the dowel rods and their shadows. [Every time 10 cm is added to the height of the dowel rod, a certain amount is added to the shadow.]
24. When the students are able to communicate that they understand that every time the stick's height is increased by 10 cm, the shadow length increases a given amount, ask the students to explain how they could use this knowledge to determine the height of the tall object since they know the length of its shadow. [Divide the shadow's length by the *average change in shadow length* calculated on the activity sheet to determine the number of increases made in height. Multiply the number of increases by 10 cm to determine the height of the tall object.]

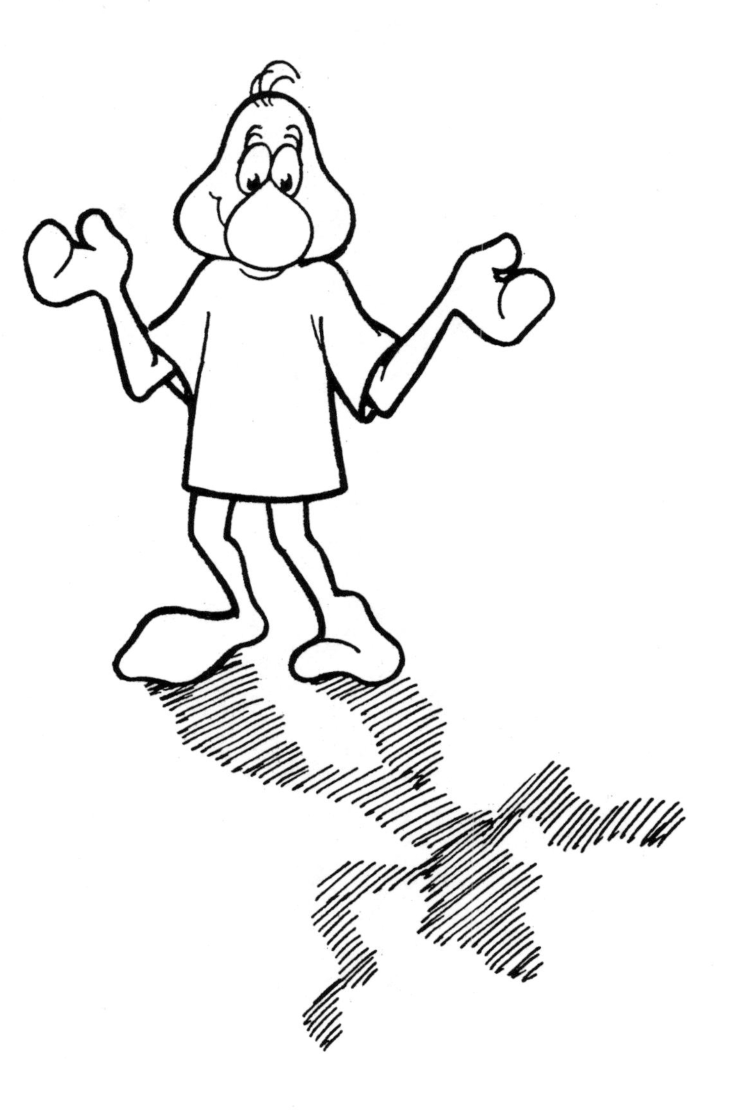

Discussion

1. How would you expect the ratio to change if you went outside and did this activity again?
2. If you were given a stick and a tape measure, how would you measure a very tall pole?

Extensions

1. Go out at different times of the day and see if there is any pattern to the ratios. [The largest height to shadow ratio will be at solar noon with the ratio decreasing in value more rapidly as you get farther away from noon.]
2. Give students a stick and tape measure and have them apply what they have learned by measuring the heights of different objects using shadows.

SIZING UP SHADOWS

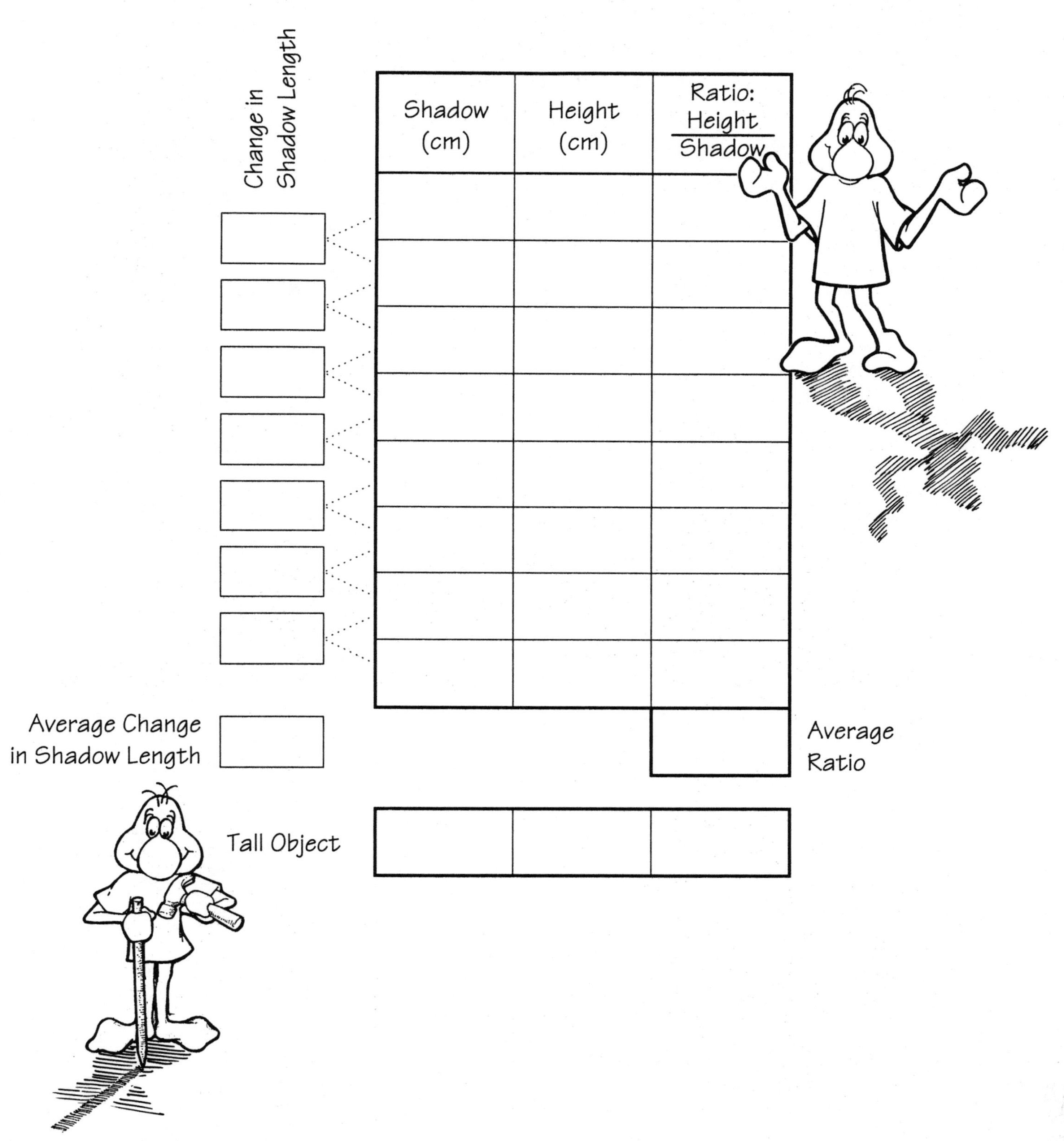

Shadow (cm)	Height (cm)	Ratio: Height / Shadow

SIZING UP SHADOWS

Height of Stick (cm)

100
90
80
70
60
50
40
30
20
10
0

0 10 20 30 40 50 60 70 80 90 100 110 120 130 140 150 160 170 180

Shadow Length (cm)

PARING IT DOWN TO SIZE

Topic
Indirect measurement

Key Question
How does the size something appears relate to how far away it is?

Focus
Students construct similar proportional characters and arrange them so they are the same apparent size. From their observations, students will develop a concept of proportionality.

Guiding Documents
NCTM Standards

- *Explore problems and describe results using graphical, numerical, physical, algebraic, and verbal mathematical models or representations*
- *Understand, represent, and use numbers in a variety of equivalent forms (integer, fraction, decimal, percent, exponential, and scientific notation) in real-world and mathematical problem situations*
- *Understand and apply ratios, proportions, and percents in a wide variety of situations*

Project 2061 Benchmarks

- *The graphic display of numbers may help to show patterns such as trends, varying rates of change, gaps, or clusters. Such patterns sometimes can be used to make predictions about the phenomena being graphed.*
- *Use, interpret, and compare numbers in several equivalent forms such as integers, fractions, decimals, and percents*
- *Organize information in simple tables and graphs and identify relationships they reveal.*

Math
Proportional reasoning
Patterns and functions
Graphing
Measurement
  linear

Science
Physical science
  line of sight

Integrated Processes
Observing
Collecting and recording data
Comparing and contrasting
Generalizing
Applying

Materials
Tagboard
Meter sticks
Tape
2-cm cubes
Scissors

Background Information

Two objects of different sizes may appear to be the same size. This similarity in the objects' apparent size is caused by the objects being different distances from the eye. The farther something is from the viewer, the smaller it will appear. An object's apparent size is directly proportional to its distance from the viewer. If an object is moved so that it is twice as far from the observer as it was originally, it will appear to be half its original size. Similarly, an object three times as tall as another object will appear to be the same size if the larger object is three times as far away as the smaller object.

This proportion makes it possible to make indirect measurements. If two objects are the same apparent size, the ratio of the distances of the objects from the viewer and the ratio of the heights of the objects will be equal. Take the example of a 9-cm high action figure which is 25-cm from your eye appearing the same height as a 180-cm tall man. The distance to the man is found using the following reasoning:

- The man is 20 times taller than the figure (180 cm ÷ 9 cm = 20).
- Since the man is 20 times as tall but appears the same size, he must be 20 times farther away. The man is 20 times farther than 25 cm. The man is 500 cm away.
- The proportion would look like this:

$$\frac{180\text{ cm}}{9\text{ cm}} = \frac{x}{25\text{ cm}};\quad x = 500\text{ cm}$$

The comparison of objects is done by linear dimensions. If two figures are similar and one is three times as wide, the larger figure is said to have a scale factor of three when compared to the smaller figure. Inversely, the smaller figure is said to have a one-third scale factor of the larger figure. It is one-third the size in all linear dimensions.

The scale factor describes the change in linear dimensions. The area changes by the square of the scale factor. A figure that has a scale factor of three will have nine times the area ($3^2 = 9$).

Management

1. Allow two forty-five minute periods for this activity. The first period is used for students to draw the different figures and the second to sight, record, and analyze the data.
2. This activity is excellent with partners if equipment allows, but can be done by groups of four.
3. Copy original figure and grid onto tagboard before beginning the activity.

Procedure

1. Distribute scissors and two tagboard copies of the grid and figure.
2. Have students determine and draw three reproductions of the figure on the grid by making one reproduction half the size of the original (scale factor 0.5, Figure 0.5), another one-and-a-half the original size (scale factor 1.5, Figure 1.5), and one twice the size (scale factor 2.0, Figure 2.0). Allow students to problem solve and discuss how they determined the size of the reproductions.
3. When students are confident they have the reproductions scaled correctly, have them cut them out along with the original (scale factor 1.0, Figure 1).
4. Have students stand the figures up and tape a cube to the back of each of the four figure's legs so they remain in the vertical position.
5. Direct the students to take a 2 cm x 5 cm strip of tagboard and fold it in half so it is 2 cm x 2.5 cm.
6. Have the students cut out a small wedge from the crease of the strip.
7. Instruct students to set the meter stick on the table. Have them tape one half of the strip to the zero end of the meter stick with the crease along the end. The other half should be perpendicular to the meter stick so that as one looks through the wedge, the rest of the meter stick can be viewed.

8. Have the students place the original figure (Figure 1) on the 40-cm mark on the meter stick.
9. Taking the figure drawn with a scale factor of 0.5 (Figure 0.5), have students move it back and forth on the meter stick until it appears to be the same size as the original figure (Figure 1) as viewed through the wedge-shaped opening in the strip. Encourage them to keep their eyes as close to the end of the meter sticks as possible.
10. Following a similar procedure, have students place the other two figures (Figures 1.5 and 2.0) on the meter stick until the figures are the same apparent size. When the students are done, the figures should be evenly spaced on the meter stick.
11. Have the students record the distance each figure is from their eyes and the height of each figure.
12. Lead students in a discussion about what patterns they see on the chart. [The distance goes up 20 cm each time. Each figure is 4 cm taller than the preceding one.]
13. Have students discuss how they could know where to place another figure of a different height if they could not move it back and forth. [height of new figure ÷ 8 = original figure distance; original figure distance x 40 cm = height of new figure]
14. Have students calculate and record the ratio and its decimal equivalent.
15. Have students discuss what the decimal equivalent tells them. [How many centimeters of height for each centimeter of distance.]
16. Direct the students to make a line graph of the data.
17. Through discussion, have students recognize that the graph is a scaled picture (map) of what they did. The sloped line they graphed is the line of sight to the top of each figure's head. To help students recognize this, they might draw a stick figure on the graph at each of the appropriate distances.
18. Ask students to find how the ratio shows up in the line. [For each centimeter of distance the line goes to the right, the line goes up the ratio's amount of centimeters.]
19. Have students follow steps 9-18 when the original figure is relocated to the 30 cm mark.
20. Guide them in a discussion of the similarities and differences in the patterns in the charts, ratios, and graphed lines.

Discussion

1. How could you use the patterns in the chart to determine the height of a new figure without measuring it?
2. How could you determine the height of a new figure using the ratio if you knew how far away it was?
3. How could you use the graph to determine the height of a new figure if you knew its distance from your eye?
4. How are the ratios different? [One is bigger.]
5. How does the difference in ratio show up on a graph? [larger ratio, steeper graph]
6. What are some experiences you have had where different-sized objects appeared to be the same size?
7. How might you use what you have learned about apparent size to measure an object of unknown height?

Extension

If students have determined a way to measure objects using apparent size, have them experiment and see if their procedure works.

PARING IT DOWN TO SIZE

Put the original man at 40 centimeters. Make the other people appear to be the same size when looking at them from the end of the meter stick. Record the measurements of these positions on the charts and graph the data. Do the same thing when the original man is at 30 centimeters.

Scale Factor	Distance from Eye	Height of Man
.5		
1	40 cm	8 cm
1.5		
2		

Height / Distance	Decimal Equivalent

Scale Factor	Distance from Eye	Height of Man
.5		
1	30 cm	8 cm
1.5		
2		

Height / Distance	Decimal Equivalent

Height of Man (cm)
0
5
10
15
0 10 20 30 40 50 60 70 80 90 100
Distance from Eye (cm)

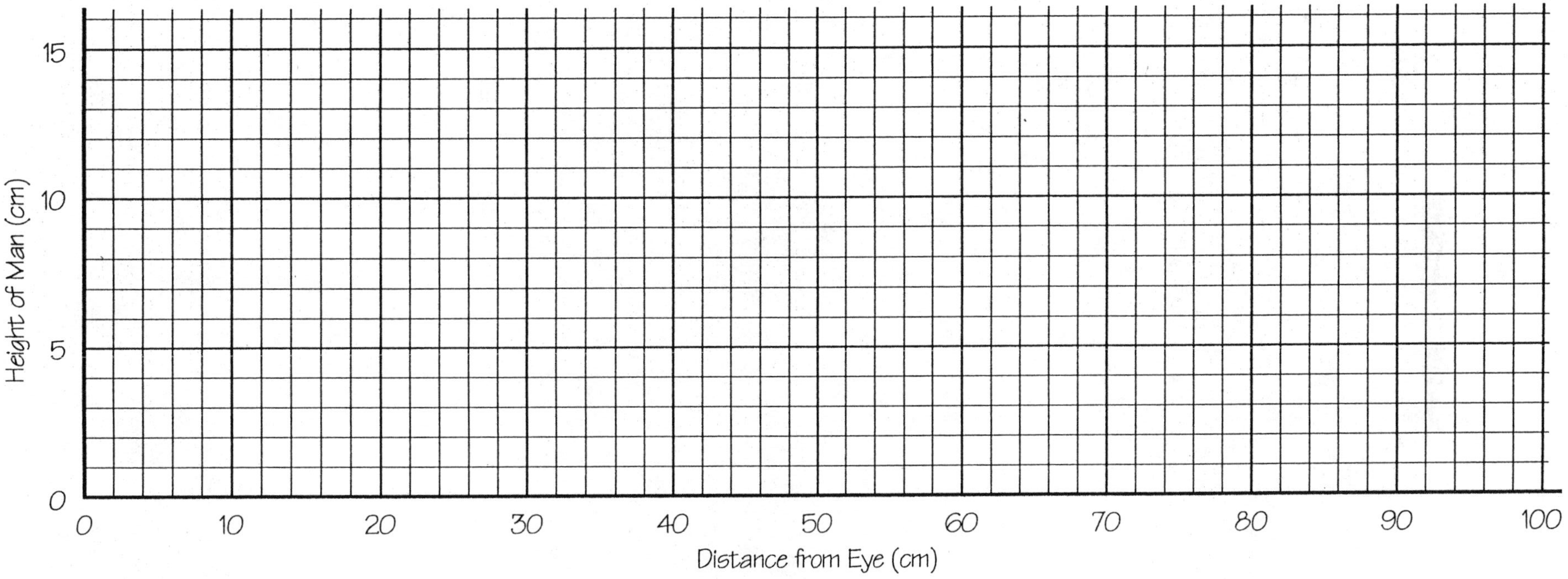

Topic
Linear and area growth

Key Question
How much of your view is blocked by an object?

Focus
Students view a rectangle and see how much of their field of vision is blocked. From their observations, students will develop a concept of proportionality as well as see the relationship of linear and area growth.

Guiding Documents
NCTM Standards

- *Explore problems and describe results using graphical, numerical, physical, algebraic, and verbal mathematical models or representations*
- *Verify and interpret results with respect to the original problem situation*
- *Understand and apply ratios, proportions, and percents in a wide variety of situations*

Project 2061 Benchmarks

- *The graphic display of numbers may help to show patterns such as trends, varying rates of change, gaps, or clusters. Such patterns sometimes can be used to make predictions about the phenomena being graphed.*
- *Graphs can show a variety of possible relationships between two variables. As one variable increases uniformly, the other may do one of the following: always keep the same proportion to the first, increase or decrease steadily, increase or decrease faster and faster, get closer and closer to some limiting value, reach some intermediate maximum or minimum, alternately increase and decrease indefinitely, increase and decrease in steps, or do something different from any of these.*
- *Organize information in simple tables and graphs and identify relationships they reveal.*

Math
Proportional reasoning
Patterns and functions
Graphing
Measurement
  linear
  area

Science
Physical science
  line of sight
  inverse-square law

Integrated Processes
Observing
Collecting and recording data
Comparing and contrasting
Generalizing
Applying

Materials
Tagboard
Meter sticks
Tape
Scissors

Background Information

An opaque object blocks the view of what is behind it. The dimensions of the area blocked are proportional to the distance the area and the object are from the viewer. An area three times as far from the viewer as the object will be three times as wide and tall as the blocking object. The area of the blocked region will be nine times as large as the blocking object.

In talking about how much bigger one thing is than another, it is important to discuss what type of dimension is being discussed. If a square two centimeters on a side is compared to another square one centimeter on a side, it is said to be twice as big. In this case, the linear dimension of length is being considered. The area of the two centimeter square is four and the one centimeter square has an area of one. The area has grown by a factor of four, while the linear dimensions have grown by a factor of two. This is because area is a two dimensional measurement. The dimension of length has doubled and the dimension of width has doubled, so the dimension of area has quadrupled, or doubled and doubled. So, how much bigger is the two centimeter square? It is twice as big in linear dimensions and four times as big in area dimensions. The area grows at a rate that is square to the linear rate of growth.

This activity provides a model of the inverse-square law. When two similar figures with one twice the size of the other in linear dimensions are lined up on a meter stick, the larger one will be twice as far from the eye as the smaller one. This is analogous to what happens when a flashlight is turned on. The light rays from the flashlight are like the lines of sight radiating from the sighting hole. If one were to measure the area covered by light at one meter and the area cover at two meters, the area twice as far away will have four times the area. Inversely, the same amount of light will cover four times the area, and the light will appear one fourth as bright. This is the inverse-squares law, a very pervasive concept in physics. What happens with light also applies to gravity, radiation, and sound.

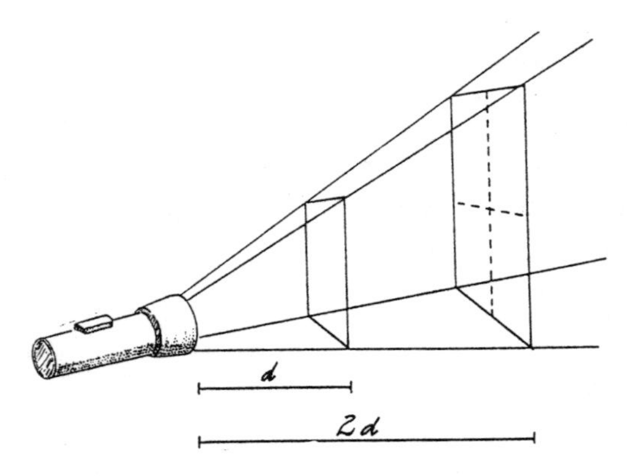

For more discussion refer to the *Background Information* in *Paring It Down to Size.*

Management
1. Allow an hour for this activity.
2. This activity is excellent in pairs, but can be done well by groups of four.
3. Copy rectangle and grid onto tagboard before beginning the activity.
4. This activity works well as an assessment. When asking students to draw the blocked region at different distances, have them discuss how they came to their conclusion. This will provide insight into their understandings of proportional reasoning.

Procedure
1. Distribute meter stick, scissors and tagboard copies of the grid and rectangle.
2. Have students take a 2 cm x 5 cm strip of tagboard and fold it in half so it is 2 cm x 2.5 cm.
3. Direct them to cut a small wedge out of the crease of the strip.
4. Instruct students to set the meter stick on the table. Have them tape one half of the strip to the zero end of the meter stick with the crease along the end. The other half should be perpendicular to the meter stick so that as one looks through the wedge, the rest of the meter stick can be seen.

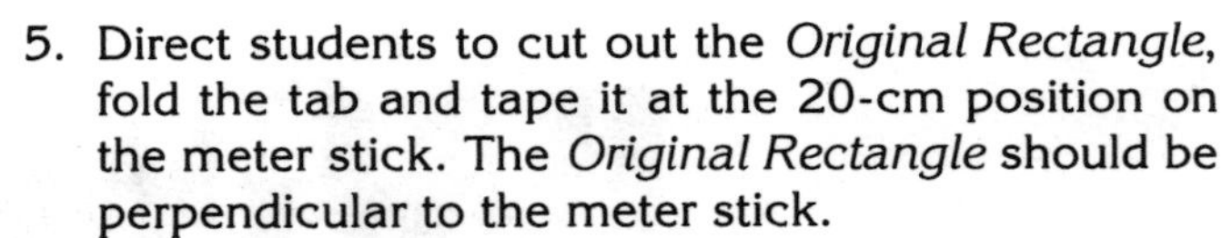

5. Direct students to cut out the *Original Rectangle,* fold the tab and tape it at the 20-cm position on the meter stick. The *Original Rectangle* should be perpendicular to the meter stick.
6. Have students predict how much of the grid will be blocked when it is placed vertically at the 40-cm mark by recording the predicted dimensions of the blocked rectangle on the chart.
7. Have one student hold the grid vertically at the 40-cm mark. A second student looks through the cutout wedge (encourage students to keep their eye as close to the end of the meter stick as possible) and identifies the edges of the region on the grid blocked by the *Original Rectangle.* The first student marks the edges of the region as instructed by the student doing the sighting. To do this accurately, have the students align a grid line with the edge of the *Original Rectangle.* The point of a pencil is placed in the blocked region on the grid. The pencil is moved slowly towards the unblocked area until the sighter sees the pencil point. Have the student mark a dot at this point. Several dots on the side and the top will identify the rectangular blocked region.
8. Using the marks on the grid, direct the students to draw the shape of the blocked rectangular region. Have them record the dimensions of the region on the chart and calculate, or count, the area of the region.
9. Follow steps 6-8, but place the grid at 60 cm and 80 cm.
10. Instruct the students to make a line graph from the data. There should be three lines (one for the width, one for the height, one for the area).
11. Have students discuss their findings and predict what would happen if the grid were placed at the 100-cm mark. Have them verify their predictions.

Discussion
1. What pattern is there in the distances chosen for the distance from the eye? [They are each 20 cm longer than the preceding one; the distances double, triple and quadruple]
2. What pattern do you find in the dimensions of the rectangles? [Each rectangle is 1 cm wider, and 3 cm taller than the smaller rectangle. The rectangles are 2, 3, and 4 times bigger in linear dimensions than the original.]
3. How do these patterns in linear dimensions show up on the graph? [lines]

4. How do the lines differ? [One is steeper.]
5. What causes a line to go up faster? [It represents the height which grows 3 cm each time, while the width only grows 1 cm each time.]
6. How many original rectangles are there in the rectangle at 40 cm (60 cm, 80 cm)? How many times bigger is the rectangle at 40 cm (60 cm, 80 cm) than the original? [4,9,16]
7. How does the area line on the graph differ from the other two? [It forms a curve.]
8. What happens to each segment of the area line as it goes to the right? [each segment is steeper than the previous]
9. Why does the area line get steeper and steeper? [The area grows at a faster and faster rate.]
10. Predict how much of the grid would be blocked at 100 cm. Explain how you came to your conclusion.

Extension

Have students consider the situation of standing near an object such as tree trunk. As one walks away from the tree, things that were blocked from view by the tree become visible. Have students try to verify this phenomena, then have them discuss and explain it. Using and exploring with the meter stick, sight, rectangle and grid may help them. [The blocked area is directly related to the ratio between the distance of the blocked region from the eye and the distance of the blocking object from the eye. As one moves back, the ratio decreases, and as a result the width of the blocked region decreases.]

	Predicted		Actual Measurements		
Distance from Eye	Width (cm)	Height (cm)	Width (cm)	Height (cm)	Area (cm^2)
20 cm			1	3	3
40 cm					
60 cm					
80 cm					

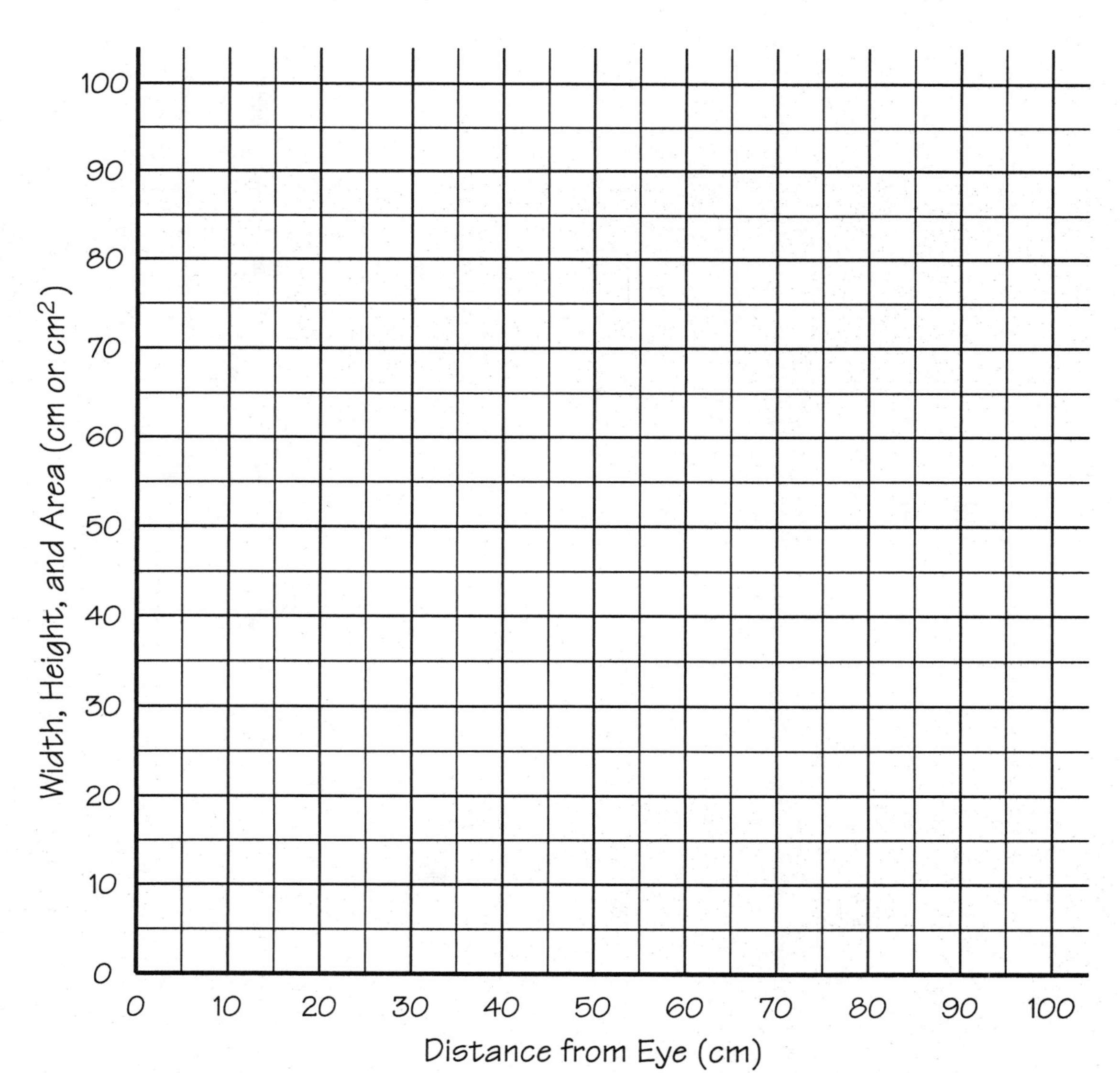

Original Rectangle

Tab

Topic
Indirect measure

Key Questions
1. When you look through a tube, how tall is the object you see through the tube?
2. If you could measure how far away you are from an object, how could you use a tube to determine the height of the object?

Focus
Students use toilet paper tubes as range finders to discover that the ratio of the view's height to the distance from the eye to the object being viewed is proportional to the tube's diameter and length.

Guiding Documents
NCTM Standards
- *Generalize solutions and strategies to new problem situations*
- *Explore problems and describe results using graphical, numerical, physical, algebraic, and verbal mathematical models or representations*
- *Extend their understanding of the process of measurement*

Project 2061 Benchmarks
- *Organize information in simple tables and graphs and identify relationships they reveal*
- *The graphic display of numbers may help to show patterns such as trends, varying rates of change, gaps, or clusters. Such patterns sometimes can be used to make predictions about the phenomena being graphed.*

Math
Proportional reasoning
Patterns and functions
Graphing
Measurement

Integrated Processes
Observing
Collecting and recording data
Comparing and contrasting
Generalizing
Applying

Materials
Rulers
Toilet paper tubes
Other tubes (from paper towel rolls or gift wrap)
Adding machine tape
Tape
Tape measure

Background Information
As one looks through a tube, the viewer finds the field of vision confined. The line of sight from the eye to the edge of the tube is the limit of the view. The lines of sight diverge as they get farther from the eye. The farther one is from an object, the greater the view. There is a direct relationship between the dimensions of the tube and the distance and width of the view. The proportion of the tube's diameter to its length is the same as the view's width at a given point to the distance to the viewer at that given point. This proportional relationship allows indirect measurements to be made.

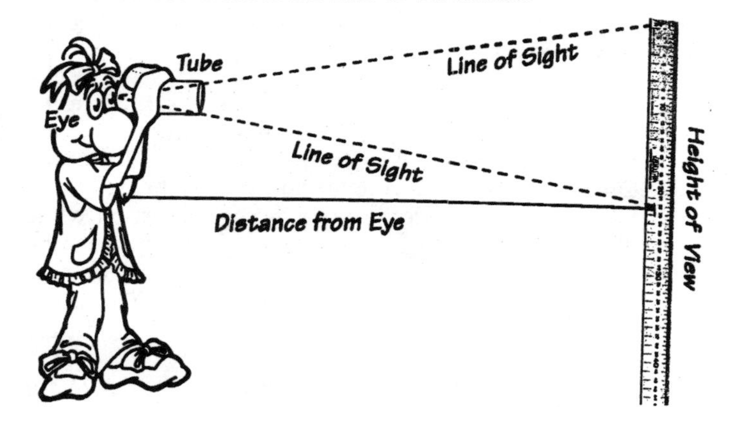

To know the height of an object with a sighting tube requires that the viewer adjust his/her position until the object appears to be the same size as the tube's diameter. From this point, the distance to the object is measured. By dividing the distance to the object by the tube length, the number of tube lengths from the eye to the object is determined. By multiplying the number of tube lengths by the interior diameter of the tube, the height of the object is determined. This process of finding the height of an object can be simplified by finding the decimal ratio of the tube diameter to the tube length and multiplying it by the distance from the eye to the object.

If one knows the height of the object and wants to know the distance the viewer is from it, the inverse ratio is used. The decimal ratio of the tube length to the tube diameter is determined and multiplied by the height of the object.

Management

1. This activity works well with pairs of students.
2. Plan on three 45-minute periods to complete this activity.
3. This activity is divided into two parts: *Part 1* has students using toilet paper tubes for sighting, *Part 2* uses various other tubes.

0 1 2 3 4 5 6 7

Procedure

Part 1

1. Distribute toilet paper tubes and have students discuss *Key Question 1.*
2. Hand out the adding machine tape, rulers, tape, and record sheets.
3. Have students take a strip of adding machine tape and use their toilet paper tube to mark off and number seven tube lengths along the strip.
4. Direct students to tape the "0" end of the numbered strip to the end of their toilet paper tube. The "1" mark will be at the other end of the tube. The remainder of the strip will extend past the end of the strip.
5. Have students tape a ruler vertically to the wall at eye level.
6. With the "0" end of the tubes up to their eyes, have students put the other end of the tube up to the ruler and record the measure they can see indicated on the ruler (to the nearest millimeter).
7. One member of each pair should hold the "2" mark of the adding machine tape next to the ruler while the viewer moves back until the adding machine tape is taut. The viewer looks through the tube noting the measure of the ruler they can see. This measure is recorded on the data sheets for *two tube lengths.*
8. Follow this same procedure (steps 6 and 7) for the remaining tube lengths.
9. Have the students discuss what patterns they see in the chart. [The height of the view increases by about the same amount each tube length.]
10. Direct the students to find the differences in height and distance between each of the tube lengths.
11. Discuss with students how much greater the height of the view gets with each tube length. Have them determine the average difference.
12. Direct them to make a line graph of the data.
13. Allow students time to complete the ratio chart and find the average ratio.
14. See *Discussion, Part 1.*

Part 2

1. Have students follow *Procedures 1-13* using tubes of different lengths.
2. Discuss how the patterns in the charts, graphs, and ratios are different for different tubes.

Discussion

Part 1

1. How could you use the average differences to determine the height of something? [Adjust your position until the object fills the view through the tube. Measure the distance to the object. Divide the distance to the object by the average tube length. The result tells how many tube lengths there are to the object. Multiply the amount of tube lengths by the average height of the view (interior diameter of the tube). The result is the height of the object.]
2. What patterns can you see in the graph? [forms an almost straight line, points are evenly spaced, the line goes up almost the same amount for each tube length]
3. What relationships exist between patterns found in the graph and the ones in the chart?
4. How could you use the graph to determine the height of an object? [extrapolate and interpolate]
5. What does the average ratio tell you about the viewing height? [how many units of height for every unit of distance, the slope]
6. How could you have found the average ratio another way? [divide the average difference in height by the average difference in distance]
7. How could you use this average ratio to determine the height of the object? [multiply the distance from the object by the ratio] Show students that the average ratio tells how many units the line goes up for every unit it goes over horizontally. This is easily demonstrated at the 10 and 100 unit marks on the horizontal (x) axis. Explain to students that this is called the *slope of the line.*

Part 2

1. How do the graphs for the two tubes differ? [different slopes, points spaced differently]
2. What differences in the tubes cause the differences in the graph? [lengths and diameters differ]
3. How do the differences in the graphs show up in the average ratio? [Different slopes show as different ratios. The smaller the ratio, the less steep the line.]
4. Describe the type of tube you would want to measure a very tall building in a crowded city and explain why. [Short tube with a large diameter. You can get a lot of view without going too far away.]
5. Can you only measure heights with a tube? Explain your answer. [You can measure any dimension to which you are perpendicular because all the diameters of a circle are congruent.]

Extension

Have students apply their measuring methods to measure the heights of different objects around the school. Have them choose which tube will be most appropriate for the measurement.

TUNNEL VISION

Difference in Distance	Distance from Eye: Centimeters	Distance from Eye: Tube Lengths	Height of View (cm)	Difference in Height	Ratio: Height / Distance	Ratio: Decimal Equivalent
		1				
		2				
		3				
		4				
		5				
		6				
		7				

Average Difference | Average Difference | Average Ratio

Difference in Distance	Distance from Eye: Centimeters	Distance from Eye: Tube Lengths	Height of View (cm)	Difference in Height	Ratio: Height / Distance	Ratio: Decimal Equivalent
		1				
		2				
		3				
		4				
		5				

Average Difference | Average Difference | Average Ratio

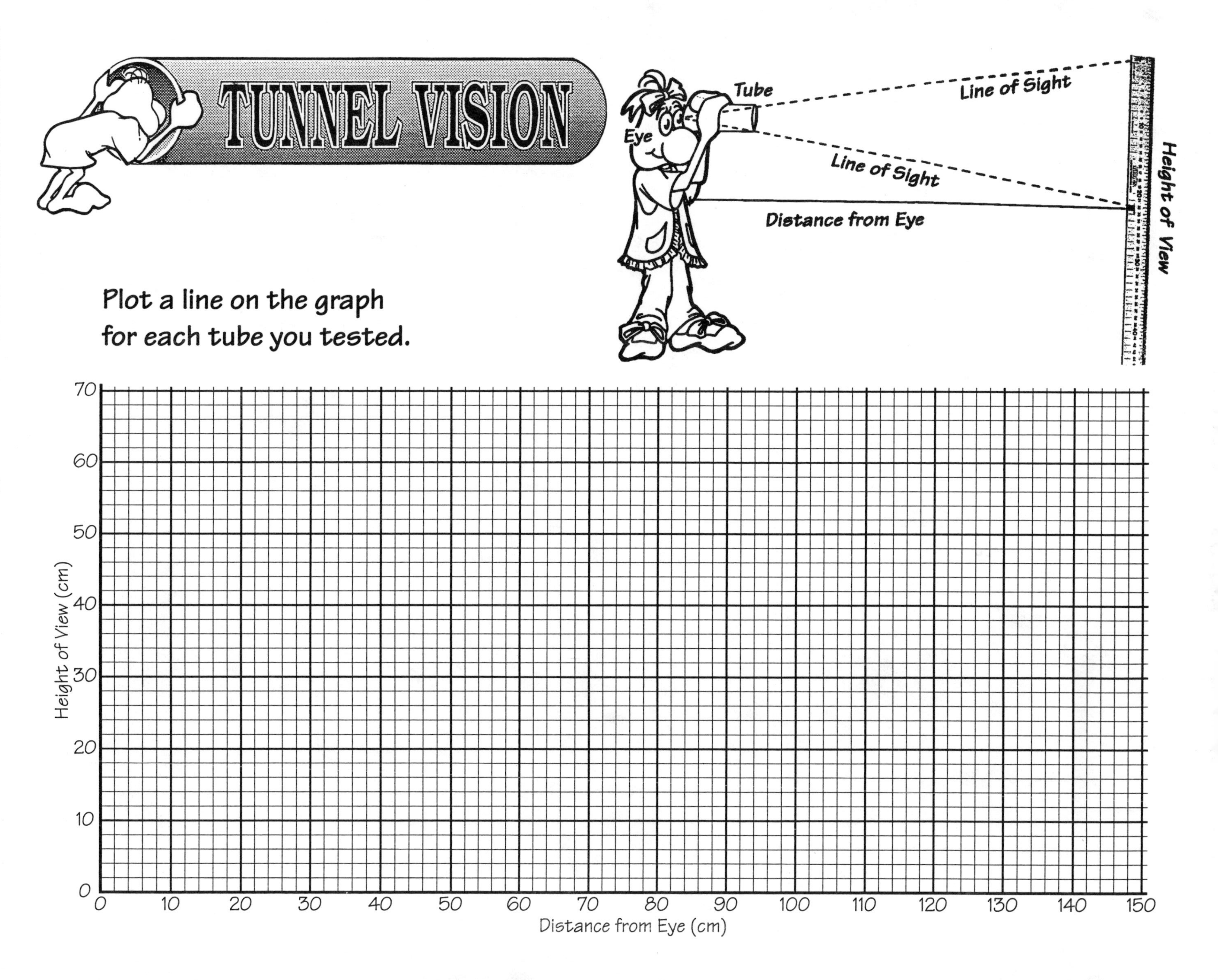
TUNNEL VISION
Tube
Eye
Line of Sight
Line of Sight
Distance from Eye
Height of View
Plot a line on the graph
for each tube you tested.
Height of View (cm)
0
10
20
30
40
50
60
70
Distance from Eye (cm)
0
10
20
30
40
50
60
70
80
90
100
110
120
130
140
150

END SIGHTS

Topic
Indirect measurement

Key Question
How could you use a sight to measure the distance from an island you are on to the shore of a river you cannot cross?

Focus
Students will make an adjustable length sight and explore the proportional relationship of the sight's height and length, to the sighted object's height and distance from the sighter.

Guiding Documents
NCTM Standards

- *Develop formulas and procedures for determining measures to solve problems*
- *Develop, analyze, and explain methods for solving proportions*
- *Use computation, estimation, and proportions to solve problems*

Project 2061 Benchmarks

- *Organize information in simple tables and graphs and identify relationships they reveal.*
- *The graphic display of numbers may help to show patterns such as trends, varying rates of change, gaps, or clusters. Such patterns sometimes can be used to make predictions about the phenomena being graphed.*

Math
Proportional reasoning
Patterns and functions
Graphing
Measurement
  linear

Integrated Processes
Observing
Collecting and recording data
Comparing and contrasting
Generalizing
Applying

Materials
Metric rulers
30-meter tapes
Meter sticks
Tape
2-cm cubes
Overhead transparencies
Scissors

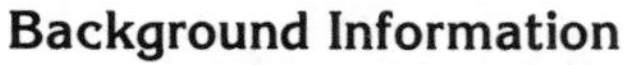

Background Information
In *Tunnel Vision,* students moved their sighting position to find the position where the tube and the object appeared to be the same size. In this activity, students adjust a sight's length to make the sight's diameter and the sighted object's height appear to be the same size.

In *Tunnel Vision,* students established the proportional relationship of a tube's diameter to length and the sighted object's height to the distance the object is from the sighter. To discover this proportionality, this activity has students compare the sighted distance to sight length and the sighted object's height to the sight's diameter.

Students will need to recognize the difference in these two proportions.

sighted height: sighted distance =
sight diameter: sight length
and
sight diameter: sighted height =
sight length: sighted length

Student will have a distinct preference for the proportion that is more intuitively comfortable for them. The critical point is that the students' understandings of how the proportions they use provide the solution for the missing measurement.

Management

1. *Tunnel Vision* should be done prior to this activity.
2. A large area is necessary for this activity. It must be at least 15 meters long.
3. Allow two to three 45-minute periods for this activity.
4. Groups of four work well for this activity.
5. Encourage groups to have all students make sightings. To get more accurate results, have students check on each other.

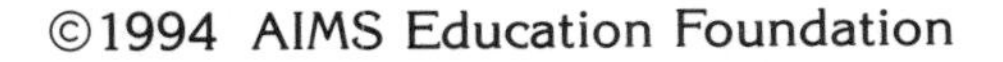

6. Instead of using tubes of various lengths, a sight will be placed on top of a ruler. The sight can be moved forward or backward to adjust the sight length.
7. The teacher may want to assemble the sights beforehand. Copy the sights from the master onto an overhead transparency. Cut around the outside of the black area. ***Do not cut out the center circle.*** Tape a 2-cm cube to the area indicated by the broken line. The sight should be perpendicular to the table top when the cube is placed on the table.

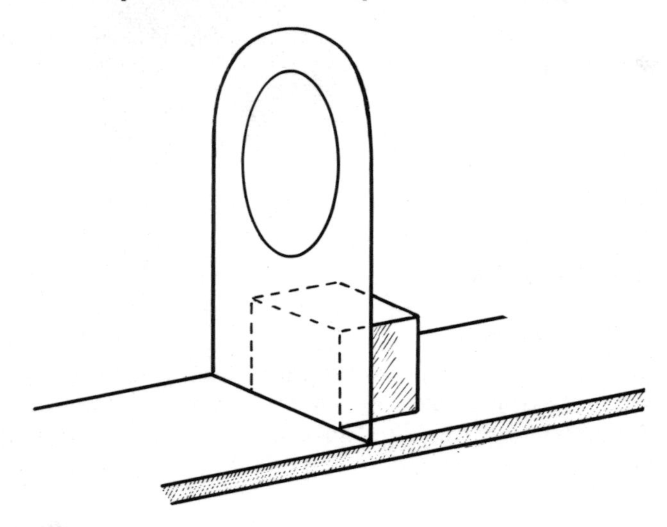

Procedure

1. Have students recall their experience in *Tunnel Vision* and discuss the *Key Question.*
2. Distribute the following materials to the students: meter stick, tape, metric ruler, 30-meter tape measure, record sheet, and 2-cm sight. (If the sights have not been assembled beforehand, have the students do it now.)
3. Have students compare the height of the meter stick to the height of the sight's circle and record it on the *Comparison of Heights* chart. [100 cm/ 2 cm = 50; The meter stick is 50 times taller than the sight's circle.]
4. Direct students to tape the meter stick vertically to a wall at eye level.
5. Have them put the end of the metric tape against the wall and extend it outward for at least 15 meters.
6. Instruct the students to put the sight at the 10-cm mark on the ruler and put the zero end of the ruler on their cheekbone just below their eye.
7. Looking at the meter stick through the circle on the sight, have them move backwards or forwards along the meter tape until the meter stick on the wall and the circle appear to be the same size.
8. After students have checked each others results and have agreed on the position where the apparent sizes are the same, have them record the distance the position is from the meter stick on the record sheet under *Sighted Distance.*
9. Direct students to move the sight to the next assigned sight length and follow steps 6-8 in a similar manner until all the assigned lengths have been completed.
10. Have students complete the chart and discuss patterns they see. [The sighted distance goes up about the same amount each time. The decimal equivalents are about the same. The ratio for the heights and distances are the same.]
12. Have students discuss what the decimal equivalent is trying to tell them. [The sighted distance is about 50 times bigger than the sight length.]
13. Have students discuss how they could use this information to determine how far they were from the meter stick. [Multiply the sight length by 50.]
14. Distribute the 4-cm sight to students and have them repeat the process.
15. Have students make a double line graph and discuss how the differences in the sights show up in the graph. [The 2-cm sight has a ratio of 50. The 4-cm sight has a ratio of 25. The ratios are the slopes of the lines on the graph. The 2-cm line is twice as steep as the 4-cm line.]

Discussion

1. How is this method better than the toilet paper tube method in *Tunnel Vision*?
2. What advantages are there for using either of the sight sizes? [As the sight gets closer to the eye, accuracy goes down. Choosing the appropriate-sized circle for the object will keep the sight farther from the eye.]
3. Which sight will let you make the most accurate measurement of a close object? [The large circle; it will be farthest from your eye.]
4. To assess their understanding, have students do some application problems such as:
   a. Your friend is 150 cm tall and is standing on the other side of a swift stream. Your friend appears to be the same size as the 2-cm sight when you hold it 29 cm from your eye. How wide is the river? [29 cm x (150 cm/2 cm) = 2175 cm = 21.75 m]
   b. You are standing 150 meters from a building. Your 4-cm sight appears to be the same size as the building when it is 25 cm from your eye. How high is the building? [150 m ÷ 25 cm = 600, 600 x 4 cm = 2400 cm = 24 m]

Extensions

1. Have students measure objects' heights and distances with their sights.
2. Have students replace the sight with a ruler and have an adjustable height and distance sight. Have them figure out how to use them to measure heights and distances.

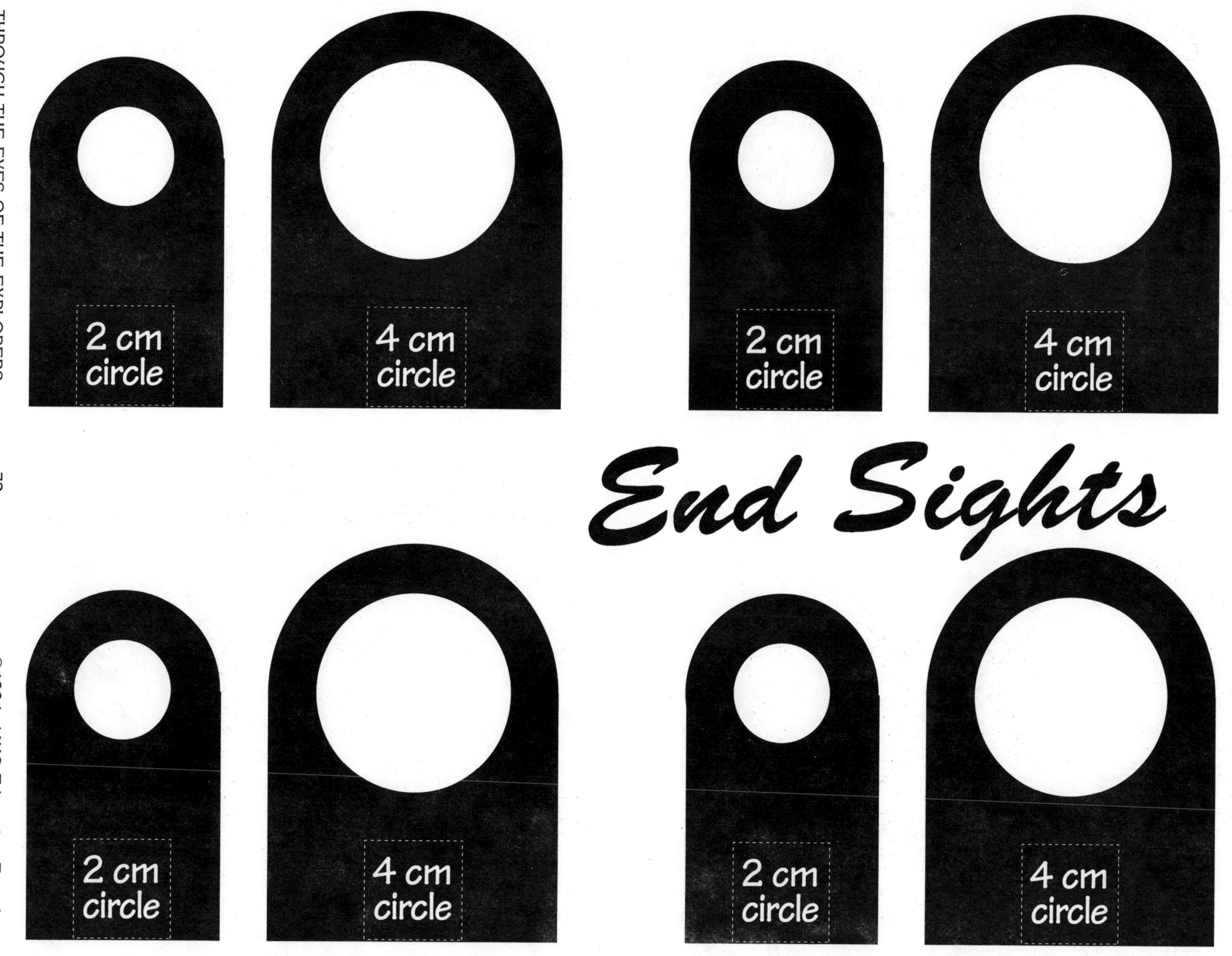
2 cm circle
4 cm circle
2 cm circle
4 cm circle
End Sights
2 cm circle
4 cm circle
2 cm circle
4 cm circle

End Sights

2 Centimeter Sighting Circle

Comparison of Diameter and Height

		Ratio	
Sight diameter	Sighted height	Sighted / Sight	Decimal equivalent
2 cm	100 cm		

Comparison of Lengths and Distances

		Ratio	
Sight length	Sighted distance	Sighted / Sight	Decimal equivalent
10 cm			
15 cm			
20 cm			
25 cm			
30 cm			

4 Centimeter Sighting Circle

Comparison of Diameter and Height

		Ratio	
Sight diameter	Sighted height	Sighted / Sight	Decimal equivalent
4 cm	100 cm		

Comparison of Lengths and Distances

		Ratio	
Sight length	Sighted distance	Sighted / Sight	Decimal equivalent
10 cm	cm		
15 cm	cm		
20 cm	cm		
25 cm	cm		
30 cm	cm		

END
SIGHTS

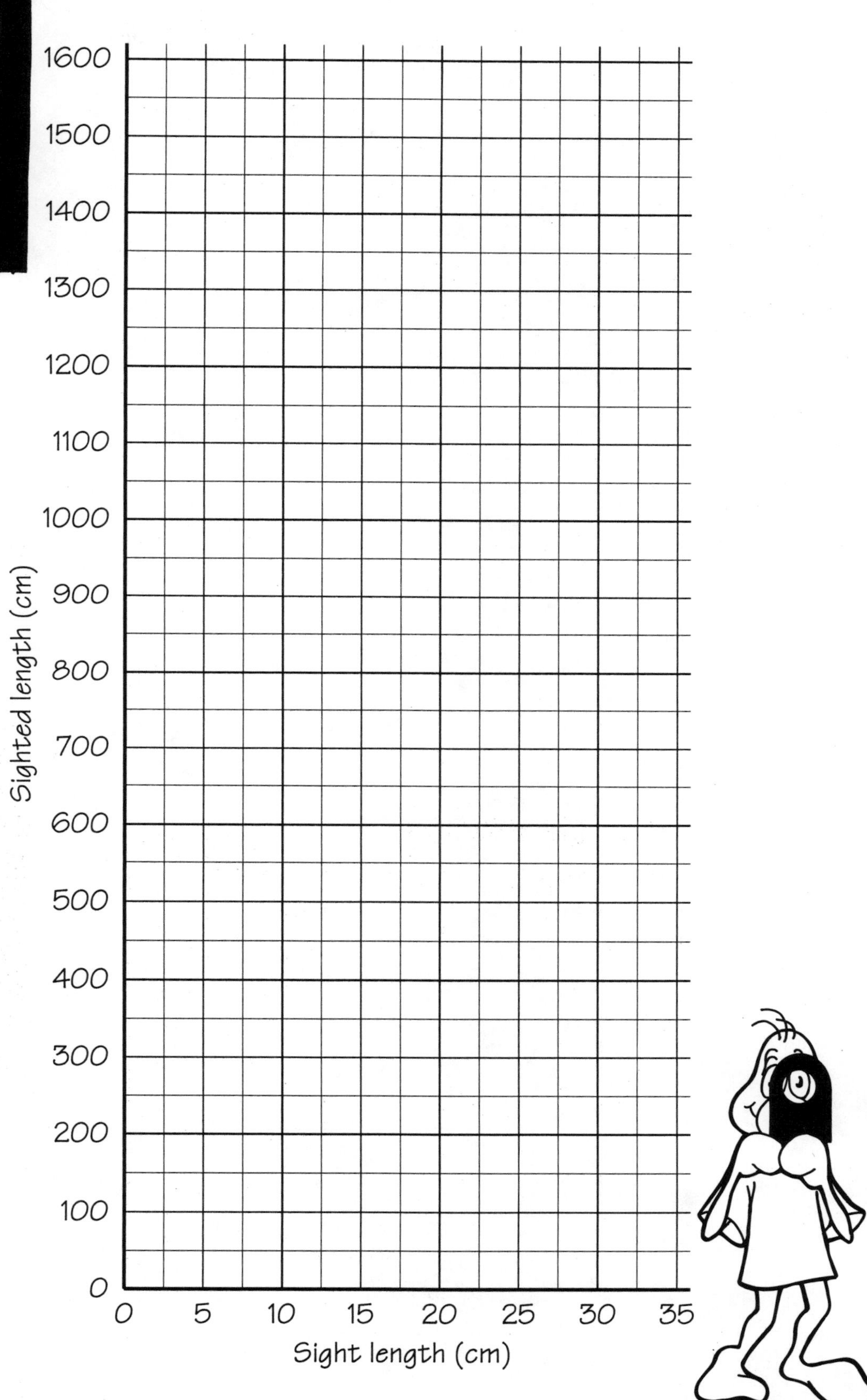
1600
1500
1400
1300
1200
1100
1000
900
800
700
600
500
400
300
200
100
0
Sighted length (cm)
0
5
10
15
20
25
30
35
Sight length (cm)

MEASURING UP

Topic
Indirect measurement

Key Question
What is the relationship between the height of an object, the distance you are from the object, and the angle of sight to the top of the object?

Focus
Students study different angles of elevation to discover that each angle has a proportion between the height of the object being measured and the sighters' distance from the object.

Guiding Documents
NCTM Standards
- *Describe and represent relationships with tables, graphs, and rules*
- *Analyze functional relationships to explain how a change in one quantity results in a change in another*
- *Use patterns and functions to represent and solve problems*

Project 2061 Benchmarks
- *Organize information in simple tables and graphs and identify relationships they reveal.*
- *Mathematical statements can be used to describe how one quantity changes when another changes. Rates of change can be computed from magnitudes and vice versa.*

Math
Proportional reasoning
Graphing
Patterns and functions
Measurement
  linear
  angular
Trigonometry

Integrated Processes
Observing
Collecting and recording data
Interpreting data
Applying

Materials
Glue
Clear packing tape
String
Pennies
Push pins
Tape
Meter stick (per group)

Background Information
Triangulation is based on the formation of similar triangles from which proportions can be made. Trigonometry is the study of the proportions formed by right triangles.

In this activity it is assumed that the object will be at right angles to the earth. When the top of the object is sighted with an angle sight, two similar right triangles are formed. The small one is on the angle sight. The large one is formed with angles at the top of the object, the sighter, and the object at eye height.

Each angle sight in this activity is drawn for only one size angle. As students sight a vertical tape measure through the straw, they will recognize that a pattern develops. A sighter with an eye height of 1.5 m using the 35 degree angle sight at one meter from the tape measure will see 2.2 m in the center of the sight. At two meters away, the sighter will see 2.9 m and at 3 meters away 3.6 m would be in the sight. Each time the sighter moved a meter farther away, the height increased by 0.7 m.

When students do this activity, many will not recognize the pattern because the numbers will not be as accurate as those in the example. However, when the data are graphed, they will produce a relatively straight line. Students will recognize that for every meter the distance increases, the height increases a given amount. The graph the students draw will be similar to the drawing on the angle sight. Students will observe that the line of best fit does not intersect at 0. As they compare their graph to the angle sight, they will realize that the height of the line at the y-intercept represents the sighter's eye height.

As students make a graph for a 45°, 35°, and 25° angle sights, they will recognize that both the graph and the sights are representations of the sightings they took. They will also recognize that they are making scale pictures of what they did. They will see the similar triangles in the angle sight and graphs and how these are similar to what happened in the real world. As students see three distinct lines on their graph, they realize that with each angle, the height increases at a different rate causing different slopes on the respective lines.

From observations made through the activity, students can be asked to develop a way to determine the height of an unknown object. One solution is to use one of the clinometers (a *clinometer* is an instrument for measuring angles of slope or inclination) and move away from the object until the top of the object is seen through the leveled clinometer. By multiplying the distance (in meters) from the object by the amount the height increases each meter for that clinometer, and adding the sighter's height, the object's height can be determined. This can be written as a function: Height = (Height difference per meter) X (Distance in meters) + (Eye height).

Some students will recognize that the angle sight is a scale drawing and can be used to determine the height of an object. First they find the scaled distance they are from the object on the horizontal axis of the sight. Then they go vertically from this point until the sighting straw is intersected. From this intersection, they go horizontally to the right until they intersect the leveling line. This is the scaled height of the object less their eye height.

Management

1. This activity is best done in groups of four: one student sights, one student levels, one student measures, and one student records.
2. Allow two periods for this activity. One period is needed to gather the data, and one period is used to graph and analyze the data.
3. Before doing this activity, the teacher needs to find an object from which to hang a long tape measure. The tape measure is secured at the ground and is held vertically so different heights can be read. The area around the tape measure must be level for at least a distance equal to the tape measure's height. A flag pole is an acceptable object to measure. The top of the tape measure can be attached to the flag pole's rope and hoisted until the tape measure is extended.
3. Prior to the activity make one 10-meter tape measure following these instructions:
   a. Make 10 copies of the tape measure on two different colors of paper.
   b. Cut each sheet into strips and use scissors to trim the tabs.
   c. Glue the strips into meter sections.
   d. Glue the meter strips together alternating colors for each meter.
   e. Number the diamonds in sequence to mark the meters.
   f. Laminate or cover the tape measure with clear packing tape to protect them.
4. Construct angle sights for each sized angle following these instructions.
   a. Copy or mount the angle sight onto tagboard or heavier cardboard.
   b. With a push pin, poke a hole in the center of the small circle in the upper right corner of the angle sight.
   c. Thread string through the hole allowing two centimeters to extend on the back side. Secure the string on the back with tape.
   d. Cut the string so that when it hangs freely, 10 cm extends below the angle sight.
   e. Tape a penny to the end of the string to serve as the weight for the plumb line.
   f. Tape a straw along the *Sighting Straw* line.

   Each group does not need a set of angle sights since the sights can be traded.

Procedure

Gathering Data

1. Have students discuss the *Key Question.*
2. Distribute the angle sights, record sheets, and meter sticks.
3. Discuss with students how these angle sights are used. The *sighter* holds the angle sight so the vertical tape measure can be seen through the straw. The *leveler* adjusts the angle of the sight until the plumb line is matched with the *Leveling Line.* The *sighter* then reads the height of the tape measure through the center of the straw.
4. Go out to where the vertical tape measure is hanging and have each sighter measure their eye height by standing by the tape measure, holding the straw parallel to the ground and sighting the tape measure height. The eye height is recorded on each chart as the *Height through straw* at a *Distance from tape measure* of 0 m.
5. The *measurer* marks a position one meter from the tape measure.
6. Using one of the angle sights, the *sighter* stands at this position and a sighting of the height of the vertical tape measure is taken and recorded.
7. A sighting is taken and recorded at each meter farther away from the tape measure until the chart is completed or the sighting is above the top of the tape measure.
8. Students trade their sights for a different-sized angle and complete steps 5 - 7 until readings with all three angle sights have been made and recorded.

Analyzing Data

9. Students find the differences between each of the adjacent *Height through Straw* and record them in the spaces provided to the right of each chart.
10. Students make a coordinate graph of the data. There will be three lines on the graph, one for each chart of data.
11. Students find patterns in the charts, graph, and angle sights and apply them to make a generalization about the *Key Question.*

Discussion

The following discussion questions focus student attention on the patterns in the chart, graph and angle sights. These three representations present the same information in different formats. A student may recognize the pattern in one format that was unclear in another. Likewise, understanding from one representation may provide understanding in another. The questions are separated into the three representations. Students will follow unique routes in assimilating knowledge. These questions provide only a sketch of what types of questions should be asked. The teacher needs to be attentive to the students' insights and appropriately order the questions for the students.

Charts

1. What patterns do you see in the charts? [The height difference is always about the same amount; with 45° the height difference is always about one; the smaller the degrees on the angle sight, the smaller the difference added.]
2. How would you predict the height of something if you saw the top of the object in the straw when you were sighting from 12 meters? [Add the difference twice to the height at 10 m. (Students may want to average the differences for each chart to have a more accurate measure of the difference.)]
3. How could you determine the height of something if you knew which angle sight you were using and the distance from the object you were when you made a sighting to the top of the object? [Add the same amount of differences to the height at 10 m as there are meters of distance past 10 meters. Multiply the distance by the difference and add the sighter's eye height.]

Graph

4. What patterns do you see in the graph? [The data produce lines. The greater the angle on the sight, the steeper the line. The lines all intersect the height axis at about the same spot above 0.]
5. What do these patterns tell you about the data? [Consistently every meter you move away, you add the same amount to the height. The greater the angle of the sight, the greater change in height. The intersection at the height axis represents the eye height of the sighter.]
6. How are the patterns on the charts similar to the patterns on the graph? [The average difference on the chart is the slope of the line on the graph (how far the line goes up for each meter of distance). The greater difference gives a greater slope which makes the line steeper. The height at a distance of 0 meters (eye height) is the same as the intersection of the graphed line on the vertical axis (the y-intercept).]
7. How could you use the graph to predict the height of an object? [interpolate or extrapolate the graph]
8. Write a function for each line.:
   general function:
   height = (slope x distance) + eye height
   45 degree function: H = (1.00 x distance) + E
   35 degree function: H = (0.70 x distance) + E
   25 degree function: H = (0.47 x distance) + E

Angle Sight

9. What similarity is there between the angle sight and the graph? [slope of sighting straw and corresponding line on graph, congruent angles on sight and graph]
10. How could you use this sight like the graph? [Go vertically from the number that represents the sighter's distance from the object. This vertical line is a scaled drawing of the object's height. Go horizontally from where the vertical line intersects the center of the sighting straw to get the scaled height of the object. The scale can be multiplied to provide a scale to measure taller objects.]
11. What will need to be added to the height on the angle sight to make it accurate? [eye height]

Extensions

1. Have groups of students measure the heights of various objects around the school. Make a bar graph of height ranges for each object and compare different groups' measurements.
2. Have students compare the angle sight method to the shadow method in *Sizing Up Shadows*.

Glue Tab Under

Glue Tab Under

Glue Tab Under

Glue Tab Under

.2

.1

.5

.4

.3

.7

.6

.5

.9

.8

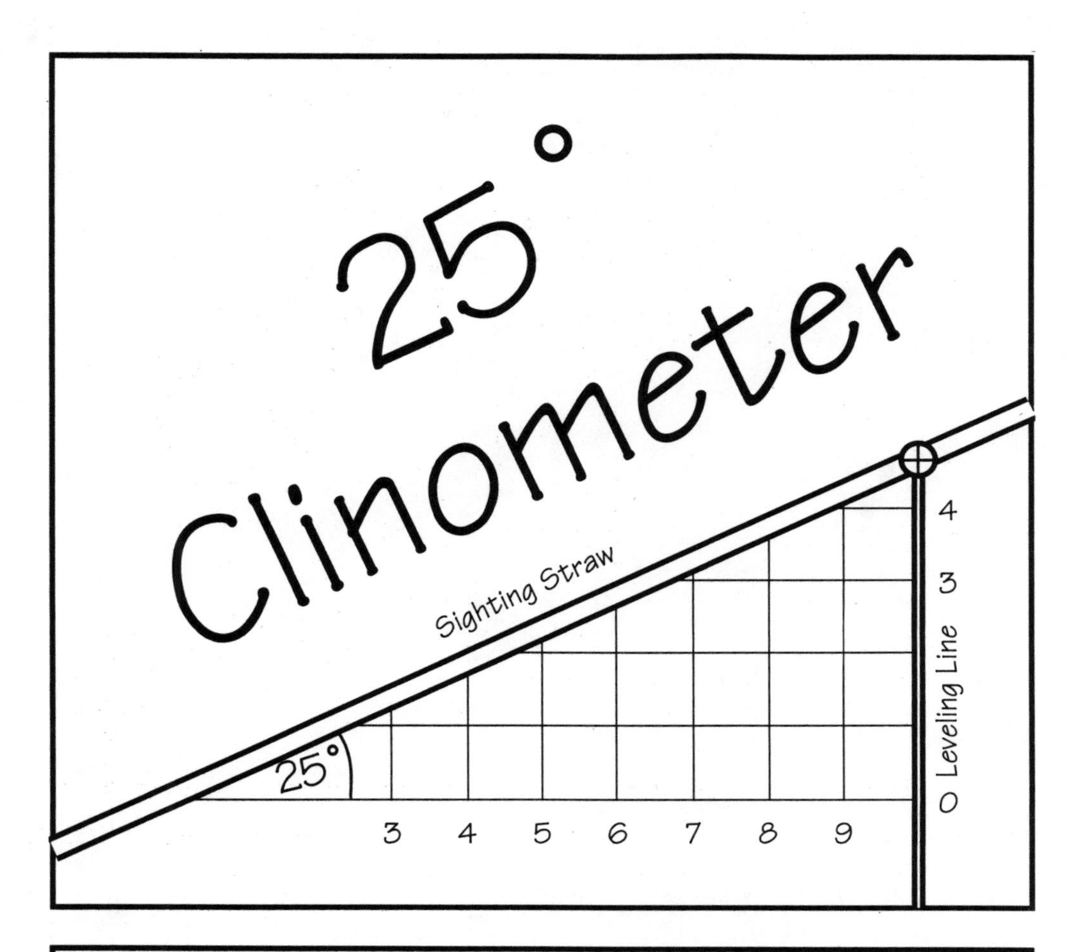
25°
Clinometer
Sighting Straw
25°
Leveling Line
0
3
4
3 4 5 6 7 8 9

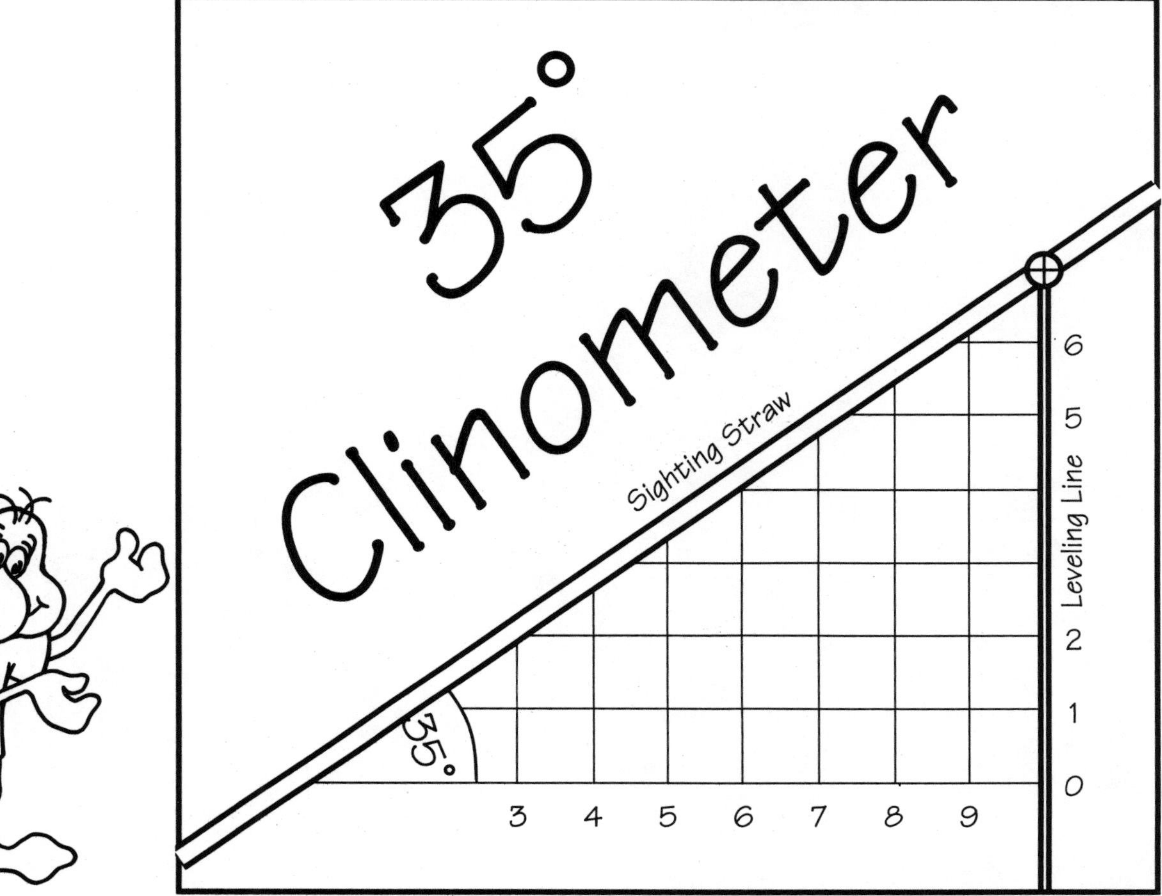
35°
Clinometer
Sighting Straw
35°
Leveling Line
0
1
2
5
6
3 4 5 6 7 8 9

45°
Clinometer
Sighting Straw
Leveling Line
45°
0
1
2
3
9
8
7
6
3
4
5
6
7
8
9
35°
Clinometer

MEASURING UP

45 Degree Chart

Distance from tape measure (m)	Height through straw (m)	Difference in Height
0	(Eye Height)	
1		
2		
3		
4		
5		
6		
7		
8		
9		
10		

D	

35 Degree Chart

Distance from tape measure (m)	Height through straw (m)	Difference in Height
0	(Eye Height)	
1		
2		
3		
4		
5		
6		
7		
8		
9		
10		

D	

25 Degree Chart

Distance from tape measure (m)	Height through straw (m)	Difference in Height
0	(Eye Height)	
1		
2		
3		
4		
5		
6		
7		
8		
9		
10		

D	

MEASURING UP

Make a line graph from your data for each of the three different clinometers.

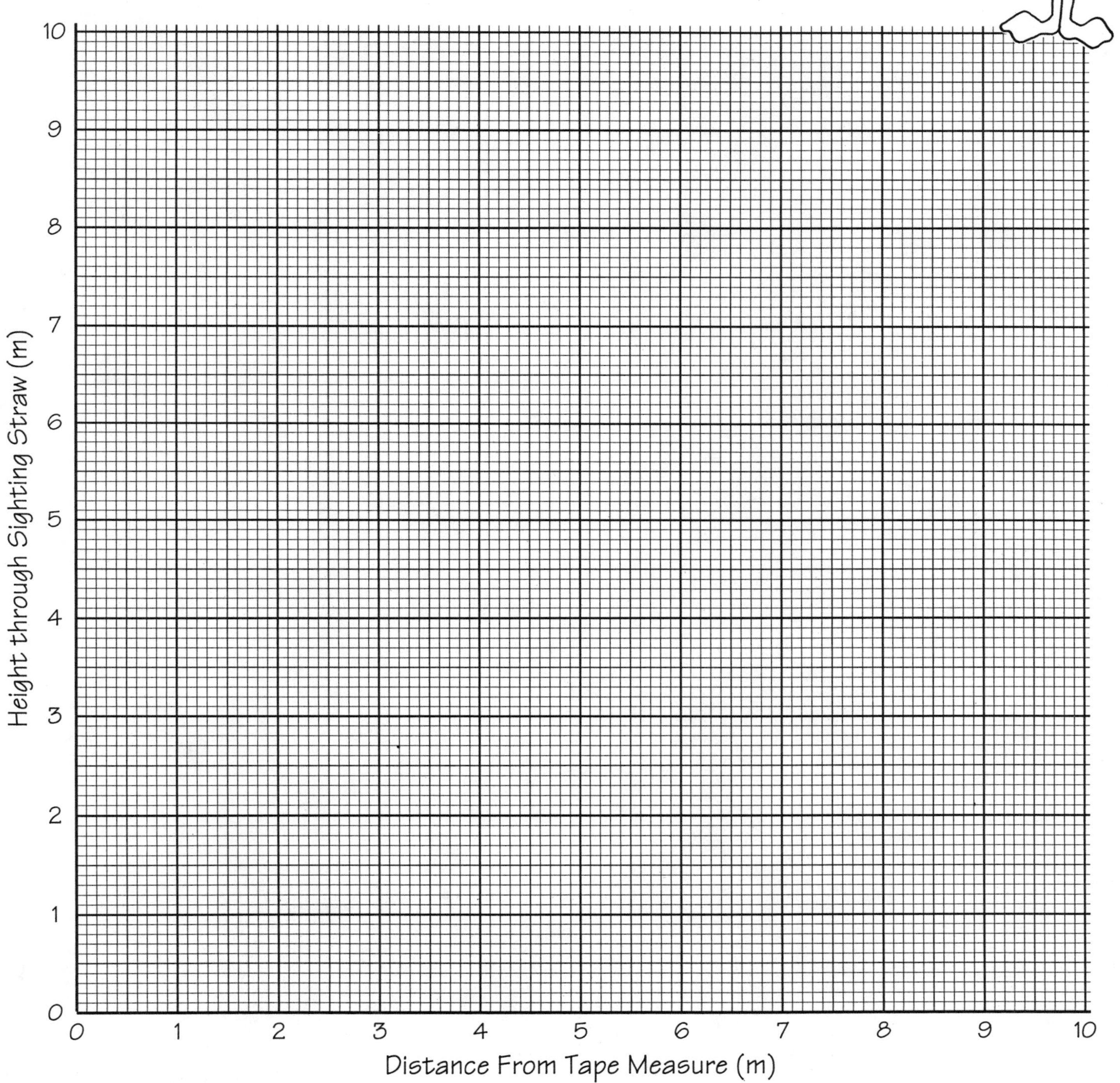

What patterns do you see in the graph?

Topic
Indirect measurement and trigonometry

Key Question
How could you measure the height of a potato projectile from a potato popper?

Focus
Students measure the maximum height of a moving object with a clinometer and use a graph to see the connection of trigonometry to simpler forms of indirect measurement.

Guiding Documents
NCTM Standards

- *Extend their understanding of the concepts of perimeter, area, volume, angle measure, capacity, and weight and mass.*
- *Understand the structure and use of systems of measurement*

Project 2061 Benchmarks

- *Some shapes have special properties. Triangular shapes tend to make structures rigid, and round shapes give the least possible boundary for a given amount of interior area. Shapes can match exactly or have the same shape in different sizes.*
- *The graphic display of numbers may help to show patterns such as trends, varying rates of change, gaps, or clusters. Such patterns sometimes can be used to make predictions about the phenomena being graphed.*

Math
Measurement
  linear
  angles
Proportional reasoning
Trigonometry
Using charts and graphs
Averaging

Science
Physical science
  action/reaction

Integrated Processes
Observing
Collecting and recording data
Interpreting data

Materials
40" 1/2"-EMT conduit
36" 1/2"-dowel
Potatoes
String
Pennies
Push pins
Tape
Metric tape measures

Background Information

A potato popper uses compressed air to launch potato plugs. It is constructed from a rigid tube. A plug of potato is pressed into each end of the tube. When one end of the tube is rapidly pushed down onto a dowel, one plug is pushed up the tube. The air between the two plugs is compressed by the rising plug. The pressure of the compressed air pushes the top plug out of the tube and it flies into the air. This is an example of the principle of action/reaction.

To determine the maximum height of the potato plug, a method to follow a moving object must be developed. In *Sizing Up Shadows,* students established that when there are two similar triangles, proportions can be made to solve unknowns. Students concluded in *Measuring Up* that right triangles which contain another angle of the same measure all have the same proportion. By determining the proportion for every angle, the height could be calculated for every angle.

Determining every proportion can be done easily by drawing a graph. A line is drawn the distance of one unit. A perpendicular line is drawn at one end. This perpendicular line is divided into fractional lengths of the unit. From the other end of the first line, angles of every degree are drawn. By following one of these angles until it intersects the perpendicular line, the proportion for the perpendicular segment to the length of the original line can be found. The numbers along the perpendicular line tell how many units high the intersection is above the right angle. Multiplying this proportional factor by the distance from the sighter to the object being measured produces the height of the object.

In trigonometry, the tangent function is the proportion in a right triangle of the leg opposite the angle being measured to the leg adjacent to the measured angle. On the drawn graph, the tangent function compares the height of the perpendicular line to the unit distance. The tangent table provides the same information as the graph. The student who understands the abstraction of the tangent table will find it is quicker and more precise.

Trigonometry assumes that objects are directly over their launch sites. Moving objects, like potato plugs, rarely rise exactly perpendicular to the ground. To deal with this discrepancy, sightings should be taken equal distance from the launch site in all directions. When the sighting angles are averaged, the mean angle will be closer to what the sighting angle would be if the launch were perpendicular.

The height calculated from angles measured with clinometers is the height of the object from eye level rather than ground level. Students may choose to solve this problem in two ways: They can lie on the ground when sighting, or they can add the sighter's eye height.

Management

1. This activity is designed to be done in conjunction with *Measuring Up* to helps students make the step to the more abstract level of trigonometry.
2. This activity is best done in groups of four: one to sight, one to measure angles, one to measure distance, and one to record data.
3. Allow two periods for this activity. The first period is used to study, practice, and understand this method of measurement; the other period is to measure the height of launched potato plugs.
4. Before beginning the activity, the clinometers should be made following these instructions.
   a. Copy or mount the clinometers onto tagboard or heavier cardboard and cut them out.
   b. With a push pin, poke a hole in the center of the small circle in the upper right corner of the clinometer.
   c. Thread a string through the hole to the back of the clinometer, letting two centimeters of string extend through to the back side. With tape, secure this section of string to the back side of the clinometer.
   d. Cut the string so that when it hangs freely there is a length of 10 cm of string extending below the clinometer.
   e. Tape a penny to the end of the string to serve as the weight for the plumb line.
   f. Tape a straw along the *Sighting Straw* line.
5. The clinometer is used by sighting the top of the object being measured through the sight (straw). The plumb line remains perpendicular to the ground while the clinometer is moved. The degree of elevation to the top of the object can be read on the clinometer while the sighting is being made. Measurements are most accurate when taken by two people; one person sights while the other reads the measurement.
6. EMT conduit is called for in the materials list because it is easily obtained at hardware stores. It is a thin steel pipe. It is sold in eight-foot lengths and is easily cut with a hack saw. If conduit is not available, any rigid tube with an interior diameter of less than 1/2" will do.
7. Before doing this activity, the teacher must gather the materials and practice shooting the potato popper. Put a potato on the ground and push the open end of the tube down on it. Do the same to the other end so there is a potato plug in each end. Push one plug in about 1/2" and, using this end of the tube, position it on top of the dowel. Now there is a vertical column with the tube above the dowel. Rapidly pull the tube down over the dowel and the top plug will shoot out of the top. Practice until the popper works successfully. The plug should easily reach a height of at least 6-7 meters.
8. If a potato popper is not available, a water rocket or thrown ball can be substituted for the moving object. You may want to use a thrown ball to let students practice using the clinometers before attempting to use the potato poppers.

Procedure

1. Explain or demonstrate the potato popper for the students. Have them discuss the *Key Question* and why they could not apply the methods they used in *Measuring Up* or *Sizing Up Shadows*.
2. Distribute the clinometers and have students discuss the differences between this clinometer and the angle sights they used in *Measuring Up*. [all angles shown; string moves and tells angle; not level position]
3. Distribute the graph and have students discuss the following questions:
   a. If the horizontal line represents the distance from the object, what does the vertical line represent? [the object's height]
   b. When you used the 45 (35,25) degree clinometer in *Measuring Up,* what happened to the height for every meter you backed up? [height grew by 1 m (0.7 m, 0.45 m)]
   c. Follow the 45 (35,25) degree line on the graph until it intersects the height line. What number is there? [1, 0.7, between 0.4 and 0.5]
   d. What do the numbers on the vertical line tell you? [how much you add in height for every unit of distance; what you multiply the distance by to get the height; how the height compares to the distance]
   e. How could you use the clinometer and the graph to determine the height of an object? [Know how far you are from the object. Use the clinometer to measure the angle to the top of the object. Using the angle, find the factor on the chart to multiply the distance to get the height.]
4. If appropriate distribute the tangent table and ask the following questions:
   a. What is the tangent for an angle of 45 (35, 25) degrees? [1.000, 0.7002, 0.4663]

b. How is the information on this table the same as the information on the graph? [It gives you the factor by which to multiply the distance to get the height.]
c. How is the information on this table different than the information on the graph? [It is not shown as a "picture." It is more precise.]

5. Have the students go out and practice measuring with the clinometer and the graph or table. (They may measure objects measured in past activities to check their accuracy.) At this point students may not remember that the sighter's eye height needs to be added onto the calculated height. Help students to recognize this error if it is not evident to them from their measurements.
6. Identify the launch site of the potato popper.
7. Have groups find sighting positions in all directions that are 10 meters from the launch site.
8. When students are ready, launch a potato plug.
9. Direct the groups to sight, measure the angle of elevation, and record it on the chart.
10. Do several launches to complete the chart.
11. Have students return to the classroom and share data.
12. Direct them to find the total and average angles for each launch. The average angle is used to calculate the height for each launch.

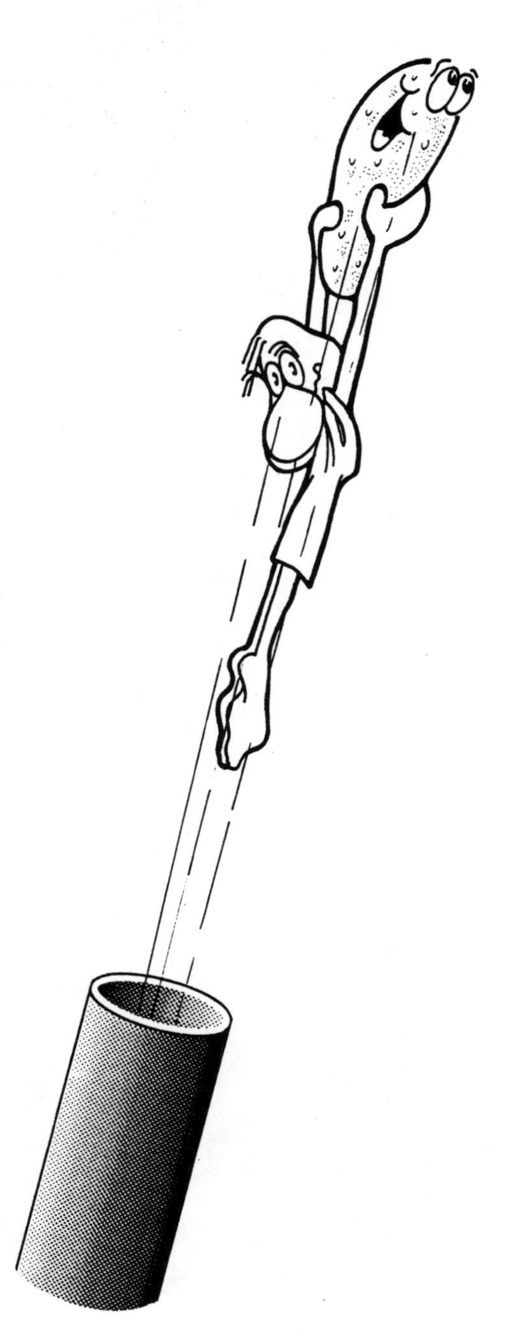

Discussion

1. What is the value of using this measuring method over the one in *Measuring Up?* [We can measure the heights of moving objects. We can stand at any distance from object; we don't have to move.]
2. Why would you not want to stand too close to the launch sight? [at large angles, small errors in angular measure cause great differences in tangent value]

Extensions

1. Have students make suggestions on how to improve the launch height of the potato popper. Guide them to experiment and test their ideas by measuring the heights of launches.
2. Have students measure the heights of model rockets as they are launched.

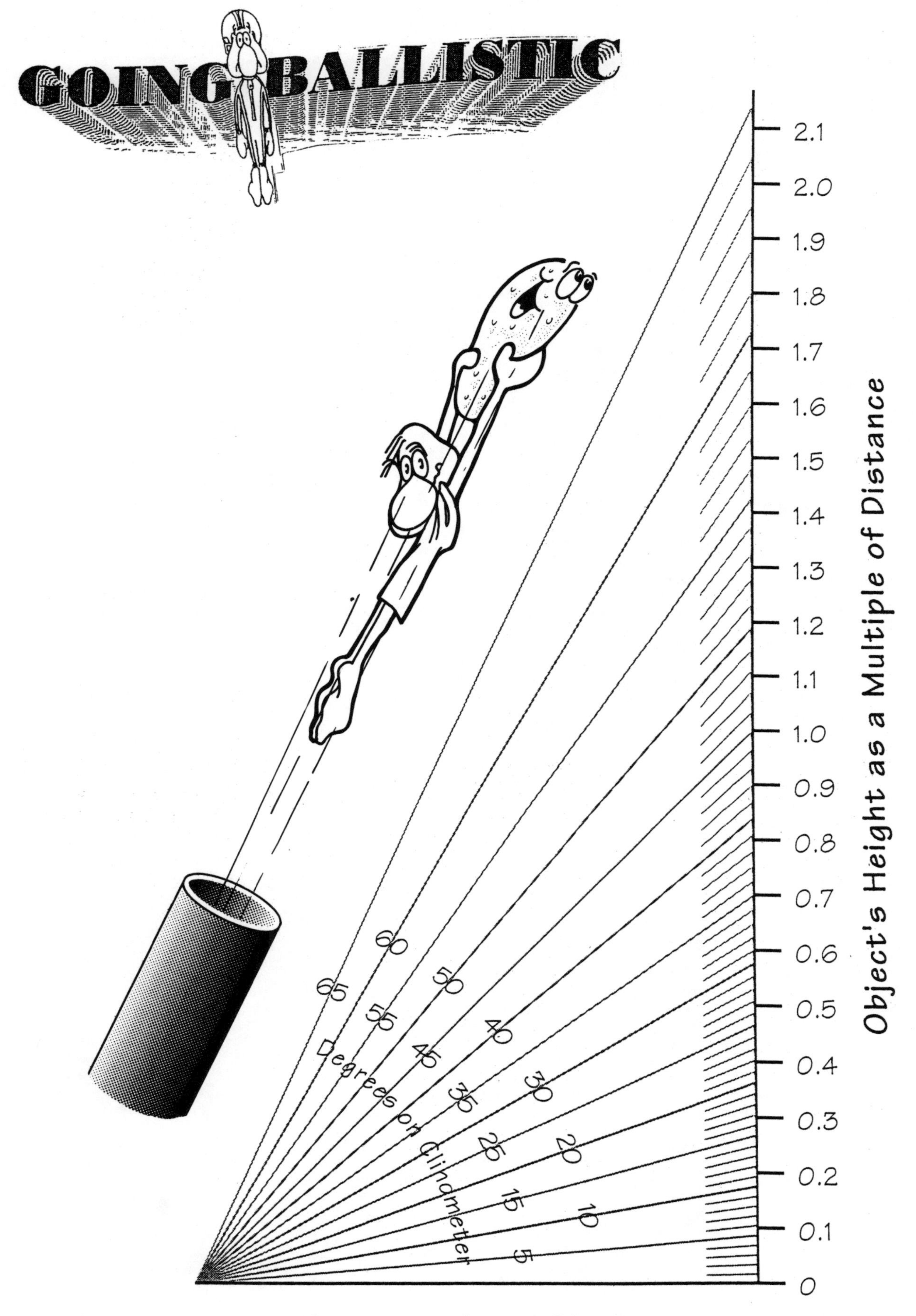
GOING BALLISTIC
2.1
2.0
1.9
1.8
1.7
1.6
1.5
1.4
1.3
1.2
1.1
1.0
0.9
0.8
0.7
0.6
0.5
0.4
0.3
0.2
0.1
0
Object's Height as a Multiple of Distance
65
60
55
50
45
40
35
30
25
20
15
10
5
Degrees on Clinometer
Distance from Object

Angle	Tangent
1	.0175
2	.0349
3	.0524
4	.0699
5	.0875
6	.1051
7	.1228
8	.1405
9	.1584
10	.1763
11	.1944
12	.2126
13	.2309
14	.2493
15	.2679
16	.2867
17	.3057
18	.3249
19	.3443
20	.3640
21	.3839
22	.4040
23	.4245
24	.4452
25	.4663
26	.4877
27	.5095
28	.5317
29	.5543
30	.5774

Angle	Tangent
31	.6009
32	.6249
33	.6494
34	.6745
35	.7002
36	.7265
37	.7536
38	.7813
39	.8098
40	.8391
41	.8693
42	.9004
43	.9325
44	.9657
45	1.000
46	1.035
47	1.072
48	1.111
49	1.150
50	1.192
51	1.235
52	1.280
53	1.327
54	1.376
55	1.428
56	1.483
57	1.540
58	1.600
59	1.664
60	1.732

Angle	Tangent
61	1.804
62	1.881
63	1.963
64	2.050
65	2.145
66	2.246
67	2.356
68	2.475
69	2.605
70	2.747
71	2.904
72	3.078
73	3.271
74	3.487
75	3.732
76	4.011
77	4.331
78	4.705
79	5.145
80	5.671
81	6.314
82	7.115
83	8.144
84	9.514
85	11.43
86	14.30
87	19.08
88	28.64
89	57.29
90	∞

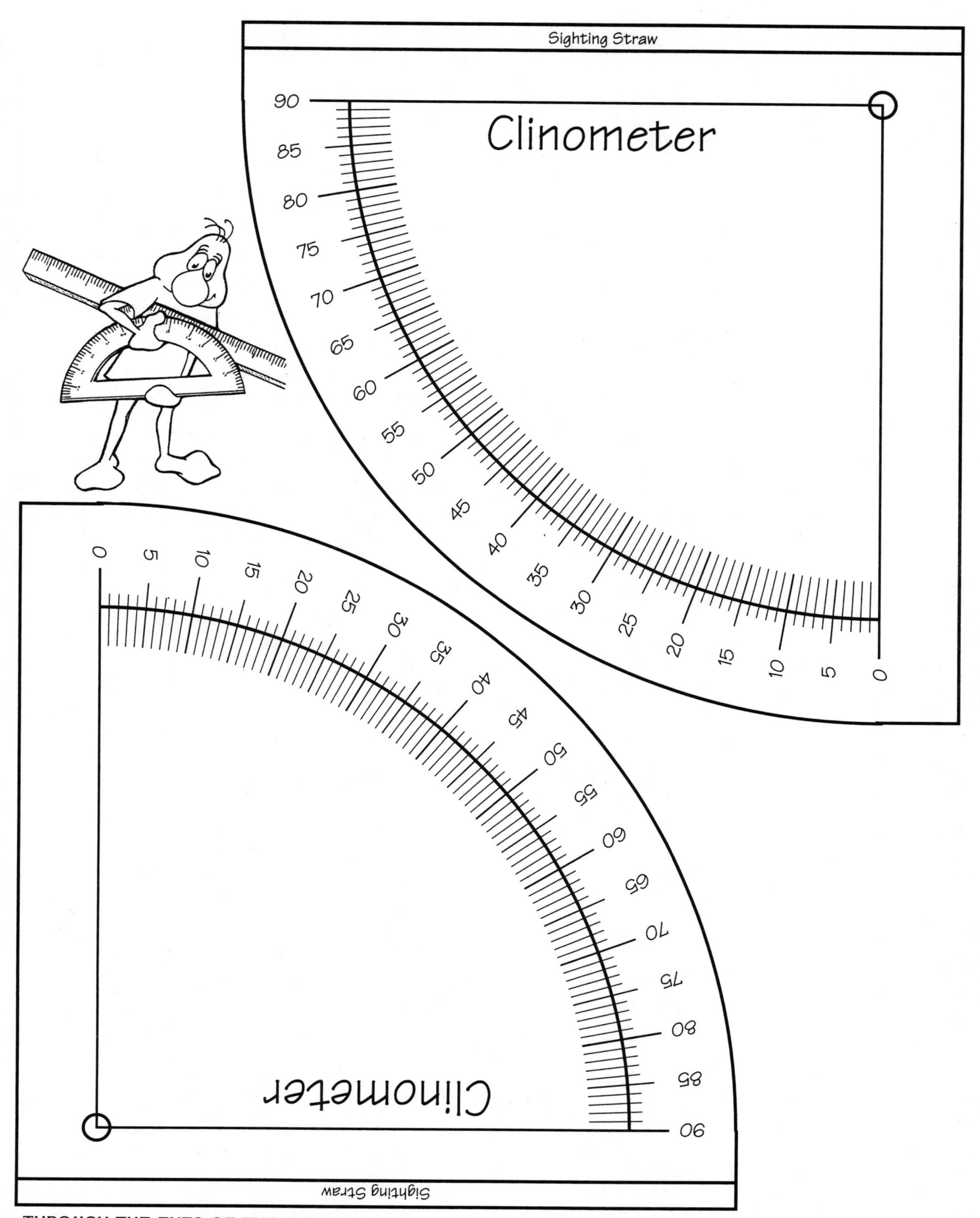
Sighting Straw
Clinometer
90
85
80
75
70
65
60
55
50
45
40
35
30
25
20
15
10
5
0
Clinometer
Sighting Straw

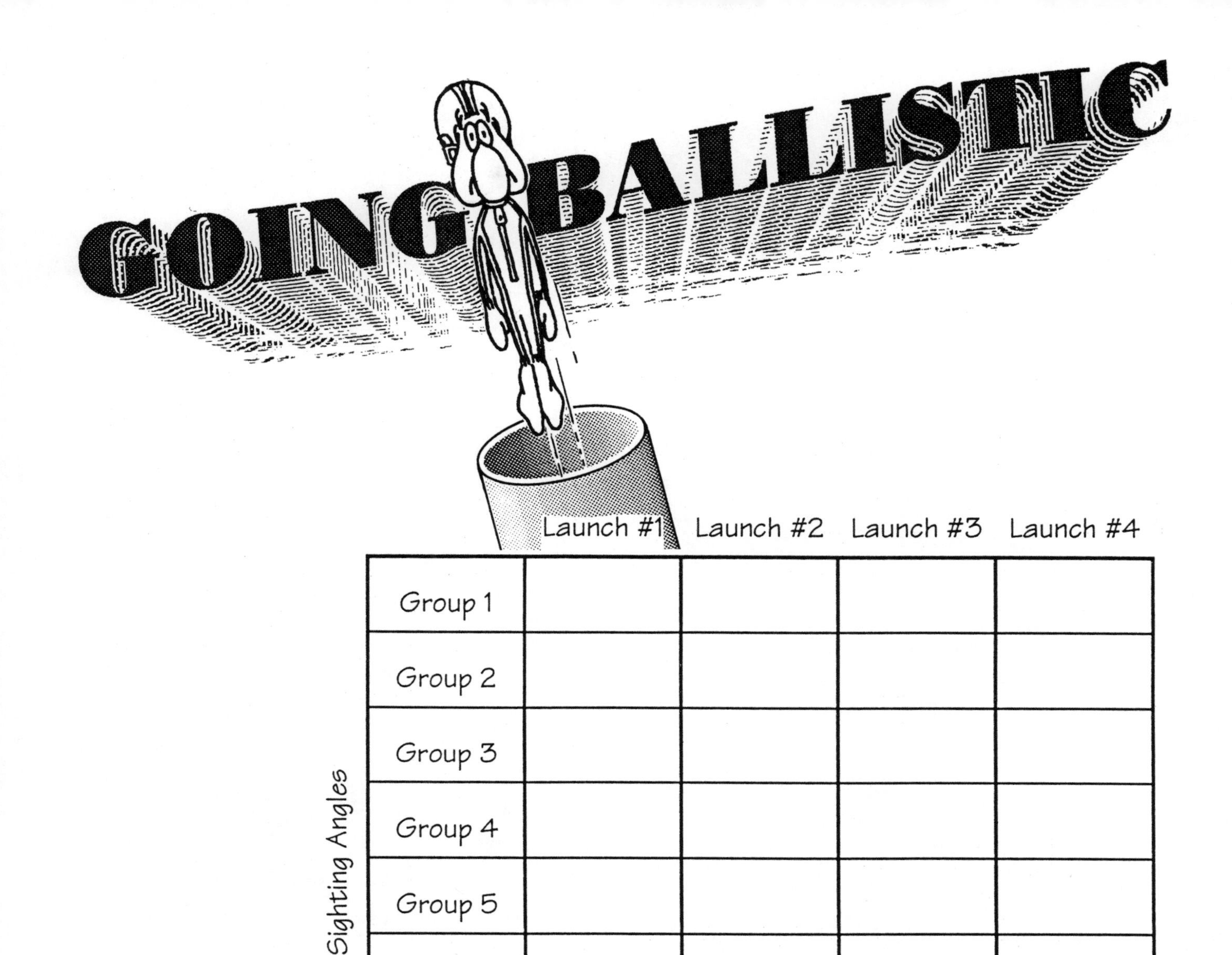

		Launch #1	Launch #2	Launch #3	Launch #4
Sighting Angles	Group 1				
	Group 2				
	Group 3				
	Group 4				
	Group 5				
	Group 6				
	Group 7				
	Group 8				
	Angle Total				
	Average Angle				
	Tangent of Angle				
	Distance from Launch				
	Calculated Height				

A MAPPING EXPEDITION

Topic
Mapping by triangulation

Key Question
How could you use the compass headings taken by Captain William Clark on his expedition to the Pacific Coast in 1805 to construct a map of the mouth of the Columbia River?

Focus
Students will construct a map of the mouth of the Columbia River from sightings recorded in journal entries by Clark.

Guiding Documents
NCTM Standards

- *Develop and apply a variety of strategies to solve problems,,with emphasis on multistep and nonroutine problems*
- *Use the skills of reading, listening, and viewing to interpret and evaluate mathematical ideas*
- *Extend their understanding of the concepts of perimeter, area, volume, angle measure, capacity, and weight and mass*

Project 2061 Benchmarks

- *Important contributions to the advancement of science, mathematics, and technology have been made by different kinds of people, in different cultures, at different times.*
- *It takes two numbers to locate a point on a map or any other flat surface. The numbers may be two perpendicular distances from a point, or an angle and a distance from a point.*
- *Estimate distances and travel times from maps and the actual size of objects from scale drawings.*

Math
Angle measure
Proportional reasoning
Scale drawings

Science
Earth science

Social Science
U.S. History
Mapping

Integrated Processes
Observing
Comparing and contrasting
Interpreting data
Applying

Materials
Protractors
Rulers
Overhead transparencies
Activity sheets

Background Information

Triangulation is the method surveyors use to determine unknown distances and precisely mark positions. If done by compass headings, the surveyor stands at a landmark and takes sightings to at least two other landmarks (*Landmarks A* and *B*). The distance traveled to *Landmark A* must be determined, then the surveyor finds the compass heading from *Landmark A* to the *Landmark B*.

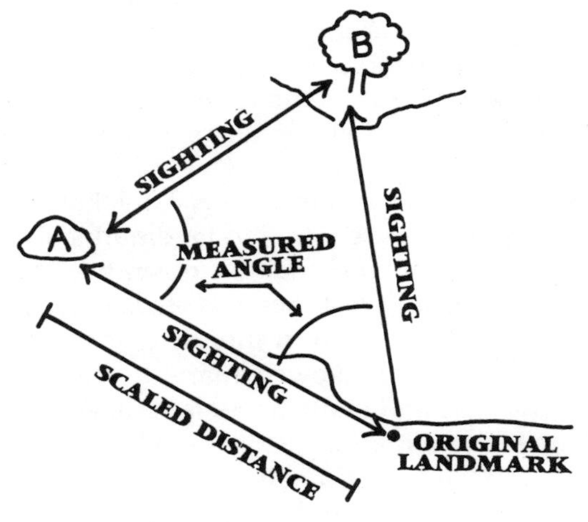

Using the headings and distance taken, the surveyor constructs a map by placing a point on the paper which represents the original landmark. The center of a circular protractor is placed on the point, and is aligned so the 0° heading is placed in the direction from the point where north will be. The degrees of headings to *Landmarks A* and *B* are measured with the protractor, and lines from the original landmark point are extended along these measured headings. These two lines represent the lines of sight. The distance to *Landmark A* is converted to the scale of the desired map. A point is placed the scaled distance from the point of the original sighting on the line of sight to *Landmark A*. This point represents *Landmark A*. The center of the protractor is placed on *Landmark A's* point and the 0° heading is aligned to north. This can be done by aligning the line of sight to this point with the heading from the original landmark. Using the protractor, a line of sight is drawn along the heading of *Landmark B*. The intersection of the line of sight from the original landmark, and the line of sight from *Landmark A* is the position of *Landmark B*.

Three points have now been drawn on the map placing them accurately by heading and scale. If the surveyor wants to know the distance from the original landmark to the *Landmark B*, it is only necessary to measure the length of the line of sight between the two points and use the scale to determine the distance.

To make a more complete map, the surveyor would take sightings to many landmarks, and would take sightings from more than two landmarks. Captain Clark took sightings from three points at up to four other landmarks. When drawing this map, some remote landmarks will have three lines of sight. The more closely the three lines come to intersecting at one point, the more accurate the sightings and the map. Captain Clark's initial sightings used in this activity give rise to a question of the accuracy of sightings from camp to Point Adams and Cape Disappointment. Changes made in a later detail map are evidence that Clark most likely realized this problem and took more sightings at a later time. The sighting to Point Adams was changed to S 41° W and the sighting to Cape Disappointment was changed to S 88° W. A discussion of the inaccuracy and remeasurement will help students to recognize the approximation and error in their sightings and measurements.

This method of map making was very quickly done from a ship. Sightings were taken to landmarks along the coastline. The ship would sail on down the coast recording its heading and distance traveled. A second set of headings were taken from this new position to the landmarks. This data allowed maps to be made which would position and scale these landmarks. The detail between the landmarks was filled in by approximating what could be seen from the ship. This method provided a relatively quick and accurate way to map the coastlines and helps to explain why there were accurate maps of coastlines well before there were maps of inland areas.

Today, compass headings are conventionally given by using 360° of heading (i.e. west = 270°). Captain Clark did not record headings in this fashion. During this time period, headings were recorded relative to north or south. A heading of S 47°E would be read "south by 47 degrees east." It might be better understood if read "47 degrees east of straight south." In conventional terms, S 47° E means a heading of 133° (180°-47°=133°). S 34° W is equivalent to a heading of 214° (180°+34°=214°).

There is only one recorded sighting taken by Captain Clark to Cape Disappointment. From this sighting, one cannot establish where on this line to place the Cape. The details of heading and distance provided in the journal can be scaled and drawn to complete the map from the camp of November 15-25, 1805 to the Cape. If the sighting, courses, and drawing have been done accurately, the point of the Cape will end up on the sight line. The distances recorded in Captain Clark's journal seem to be a little generous and the resulting maps will be slightly larger than they should be. The final map of the mouth of the Columbia River made by Clark is more accurate than ones made from these initial sightings and journal entries. The final map was a compilation of surveys and trips into the locality over the expedition's four month stay.

The journal reproduced for this activity is copied from the published edition and has only been changed in its omission of unnecessary information for the task of mapping. Students may find Clark's journal of interest because of spelling, grammar, and punctuation and may lead to an interesting discussion. Captain Clark's original journal was handwritten and then transcribed by a typesetter as accurate to the original as possible. Students may want to discuss what, if any, punctuation rules were followed, and what those might have been. Many of Captain Clark's spellings follow conventions of the time which were more like British spelling. Spellings in American English were not fixed until Noah Webster finished his dictionary in the 1820's. However, most of the words can be determined phonetically. Some abbreviations are hard to decipher. Star.d is the abbreviation for starboard, the right side of a boat as you face forward. The expedition was headed down stream towards the Pacific so the starboard side was the north bank of the Columbia. Similarly, Lar.d represents larboard, the left side or south bank of the Columbia.

Management

1. This activity is done well in pairs, one doing the reading, both discuss how to use the data, and the second drawing the map.
2. Prior to this activity make sure students are familiar with measuring angles with a protractor. Provide them with instruction on the similarities of compass headings and angles.
3. Two types of compass protractors are included in the activity pages. One has the conventional compass face. Students could benefit from the conversion to the conventional (S 34°W is equivalent to a heading of 214° because 180°+34°=214°). The other has a compass face that should help students use Clark's courses more easily. It is recommended that students use the second type if they are not familiar with compass work.
4. Copy the appropriate compass protractors onto overhead transparencies before doing this activity. Each pair of students will need a protractor.

5. Before copying the student sheet with the sighting points and the map of the Columbia, make sure the copy machine is set for 100%. Many copy machines automatically make slight reductions in their copies to ensure that all margins are copied. This reduction changes the scale of the map and will cause inaccuracies if the reduction is not overridden.

Procedure

Part 1

1. Distribute the first information sheet with the chart of sightings.
2. Have students read the information sheet and discuss the *Key Question*.
3. Distribute compass protractors, rulers, and sighting points sheet.
4. Direct students to use protractors and rulers to draw in lines of sight from each point. Make sure they are familiar with Clark's method of recording headings.
5. Have students identify intersections of lines of sight with their landmark names.

Part 2

6. Distribute information sheets with the journal to students.
7. Allow time for the students to read and discuss how the information in the journals could be made into a map. Some help and discussion may be needed in the interpretation of spellings and meanings of words. Students should recognize the compass headings and distances given in the courses as the key to the map completion.
8. Students use the courses' headings and distances to complete the map of the north shore of the Columbia River from the camp to Cape Disappointment and its adjacent Pacific Coast.
9. Have students compare their maps with each other by putting one on top of the other and holding them up to a light. Have them discuss discrepancies between their maps.
10. Distribute the map of the mouth of the Columbia River.
11. Have students compare their maps to the one drawn by Captain Clark. They should place their sighting points on top of the sighting points on the key. They should see that the intersection of the sight lines are close to the points on the opposite shore, and that their maps of the Cape approximate its shape.
12. Have students discuss what might have caused the discrepancies in the maps.

Discussion

1. How does your map differ from Capt. Clark's map?
2. What might have caused these discrepancies?
3. What evidences are there that some of Capt. Clark initial data is inaccurate? [The lines of sight do not intersect at one point. Cape Disappointment does not match its sight line.]

Extension

Get a navigational chart of an area. Have students apply their skills using compass headings with the chart.

a. Have them determine a heading and course to get from one place to another.
b. Choose a place on the map where a ship has told you it is in distress, or a treasure is buried. Measure the headings to two landmarks from this point. Give the students the headings to the landmarks and have them determine the position of the point.

Curriculum Correlation

Social Studies

Find a current map of the mouth of the Columbia River and identify its discrepancies with Captain Clark's final map.

A MAPPING EXPEDITION
(Clark's journal)

In November of 1805, the Lewis and Clark Expedition reached the mouth of the Columbia River at the Pacific Ocean. The people in the expedition made camp on a point on the north bank of the river from November 15 - 25. During their stay at this site, they took sightings to three points on the south bank and one to Cape Disappointment across a bay on the north bank of the river. Below are recorded the measurements taken from those sightings.

	Point of Rocks course N. 80°W. 1 mi.	Camp November 15 - 25, 1805	Pt. Distress course N. 78°E. 2.5 mi.
Pt. Williams		East	S. 87°E.
Pt. Meriweather	S. 50°E.	S. 47°E.	S. 29°E.
Pt. Adams	S 34°W.	S. 35°W.	S. 50°W.
Cape Disappointment		S. 86°W.	

From this information a map locating the points was made and distances could be determined. How could you use this information to construct a map of the mouth of the Columbia River?

From this base camp on the north bank of the river, Capt. Clark hiked around Haley's Bay to Cape Disappointment. He recorded his course with headings and miles traveled so that a more detailed map could be made. How could you use these journal entries to continue the map that was started with the sightings?

November 15th Friday 1805
in full view of the Ocian from Point Adams or Rond to Cape Disapointment, I could not see any Island in the mouth of this river as laid down by Vancouver. the Bay which he laies down in the mouth is imediately below me. This Bay we call Haley's Bay from a favourite trader with the Indians which they Say comes into this Bay and trades with them

November 16th Satturday 1805
The Countrey on the Star.d Side above Haleys Bay is high broken and thickley timbered on the Lar.d Side from Point Adams the countrey appears low for 15 to 20 miles back to the mountains, a pinical of which now is covered with Snow or hail, as the Opposit [shore] is too far distant to be distinguished well, I shall not attempt to describe any thing on that side at present.

November 18th Monday 1805
A little cloudy this morning I set out with 10 men and my man York to the Ocian by land.

N. 80°W. 1 mile to a point of rocks abouy 40 feet high, from the top of which the hill Side is open assend with steep assent to the tops of the mountains, a Deep nitch and two small streams above this point, then my course was

N. W. 7 Mile to the enterance of a creek at a lodge or cabin of Chinnooks passing on a wide Sand bar the bay to my left and Several Small ponds containing great numbers of water fowls to my right;

This creek appears to be nothing more than the conveyance of Several Small dreans from the high hills and the ponds on each side near its mouth. here we were set across all in one canoe by 2 squars. to each I gave a small hook

S. 79°W. 5 Miles to the mouth of Cinnnook river, passed a low bluff at 2 miles below which is the remains of huts near which place is also the remains of a whale on the

sands, the countrey low open and Slashey
S. 20°W. 4 Miles to a Small rock island in a deep nitch passed a nitch at 2 miles in which there is a dreen from Some ponds back; at 3 miles passed a nitch. this rock Island is Small and at the South of a deep bend in which the nativs inform us the Ships anchor, and from whence they receive their goods in return for their peltries and Elk skins &c.
S. 46°E. 2 Miles to the iner extremity of Cape Disapointment passing a nitch in which there is a Small rock island, a Small Stream falls into this nitch from a pond which is imediately on the Sea coast passing through a low isthmus. this Cape is an ellivated circlier point covered with thick timber on the iner Side and open grassey exposur next to the Sea and rises with a Steep assent to the hight of about 150 or 160 feet above the leavel of the water this cape as also the Shore both on the Bay & Sea coast is dark brown rock. I crossed the neck of Land low and 1/2 of a mile wide to the main Ocian, at the foot of a high open hill projecting into the ocian and about one mile in Si[r]cumfrance. I assended this hill which is covered with high corse grass. decended to the N. of it and camped.

November 19th Tuesday 1805
a cloudy rainey day proceeded up the coast which runs from my camp 1-1/4 miles west of the iner extr'y of the cape
N. 20°W. 5 miles through a rugged hilley countrey thickly timbered off the Sea coast to the Comencement of an extencive Sand beech which runs N. 10°W. to point Lewis about 20 miles distance. I proceeded up this coast 4 miles and marked my name on a low pine. and returned 3 miles back (the countrey ops^d this Sand coast is low and Slashey, crossed the point 2 miles to the bay and encamped on Chinnook river

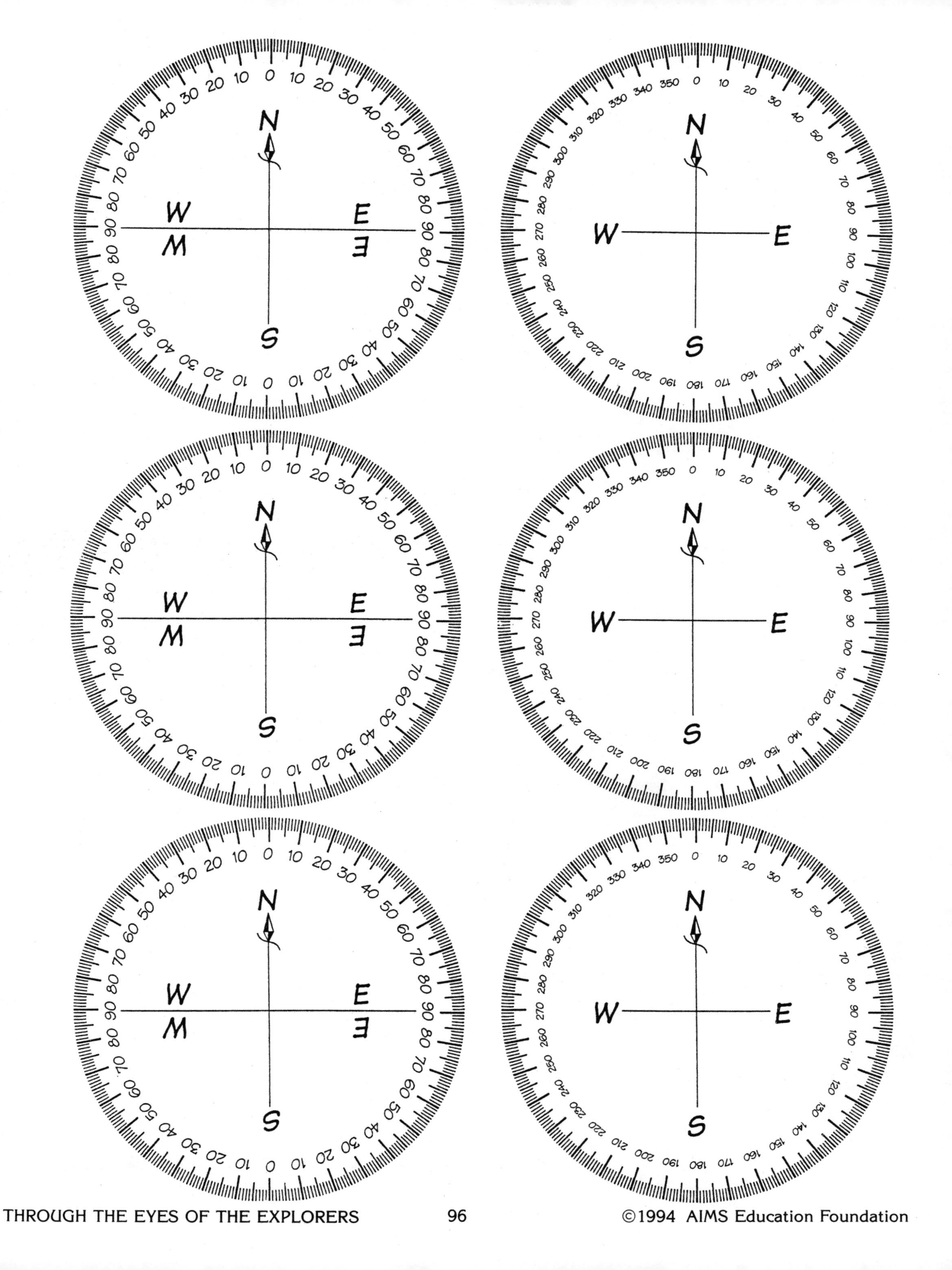

A MAPPING EXPEDITION

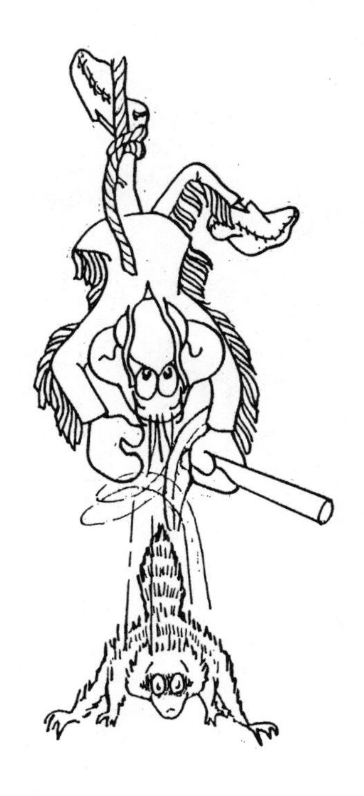

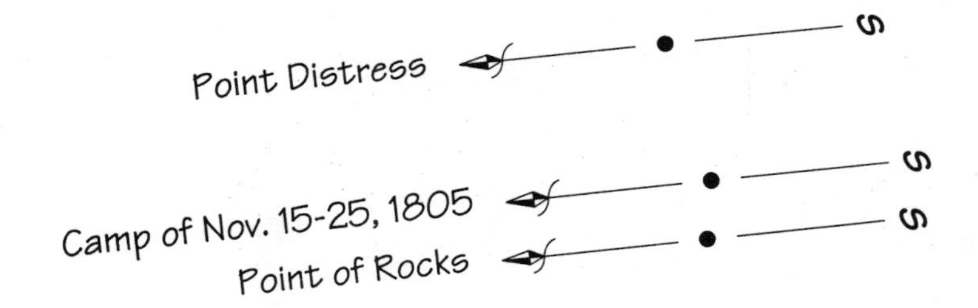

Scale: 1/4 in.= 1 mi.

MAP KEY
Cath la h mah Nation
9 large wood houses
Camped 26th Nov 1806
Camped 27, 28, 29, 30 Nov.
Point Williams
Point Meriwether
Encamped 8th 9 Nov. 1805
Encamped from the 10th thru 15th of Nov 1805
Old Chinook Village 36 H
Encamped from the 18 to 25 Nov. 1805
We-tut River
Clot sop Nation
8 large wood houses
3 Houses of Clotsops
Chin-nook Nation
1 house
Point Adams
Cape Disapointment
Mouth Of The Columbia River
Scale: 1/4" = 1 Mile

Topic
Indirect measurement by triangulation

Key Questions
1. How could you determine the distance between two places if you cannot measure it directly?
2. If you had to shoot the cannon of a ship at a target on land, how would you determine how far away the target was?

Focus
Students will use a simple transit to measure the degrees from two different places on a baseline to an object in a field. By using similar triangles, they will determine unknown distances.

Guiding Documents
NCTM Standards
- *Understand and apply geometric properties and relationships*
- *Represent and solve problems using geometric models*
- *Develop an appreciation of geometry as a means of describing the physical world*
- *Extend their understanding of the concepts of perimeter, area, volume, angle measure, capacity, and weight and mass*

Project 2061 Benchmarks
- *Important contributions to the advancement of science, mathematics, and technology have been made by different kinds of people, in different cultures, at different times.*
- *Mathematics is helpful in almost every kind of human endeavor–from laying bricks to prescribing medicine or drawing a face. In particular, mathematics has contributed to progress in science and technology for thousands of years and still continues to do so.*

Math
Geometry
triangles
constructions
Measurement
linear
angular
Proportional reasoning
Scale drawings

Social Science
Navigation
Map making

Integrated Processes
Observing
Collecting and recording data
Interpreting data
Applying

Materials
Tape measure
Rulers
Protractors
Copies of transits
Pins
Tape
Straws
Cardboard

Background Information

Triangulation is the method surveyors use to determine unknown distances and precisely mark positions. The surveyor first marks a baseline from which to measure. This baseline is often determined by a compass setting and a permanent landmark used as a reference point. Using a transit, a tool for measuring angles related to the protractor, the surveyor determines the degree of angle between the baseline and the object to which the distance is unknown. The surveyor then moves the transit down the baseline to a new sighting position. The distance the surveyor has moved is recorded. From this new position, the surveyor uses the transit to determine the degree of angle between the baseline and the same object.

With the two angles the surveyor constructs a similar triangle to the one formed by the baseline and the lines from the sighting points to the object. Since the actual triangle and the drawn triangle are similar, the sides of one have the same proportions as the sides of the other. The corresponding sides of the actual triangle and the drawn triangle have the same ratios to their corresponding baselines. By finding the ratio on the drawn triangle, the surveyor can apply the ratio to the distance between sightings on the actual baseline to determine immeasurable distances.

In this activity, the baseline is the midline of a ship running from the stern to the bow. The distance measured is from the points that represent the bow and the stern. Sightings are taken from both these positions to the target which has been chosen by the teacher.

Historically, a method using the concepts developed in this activity were used to determine the range of an enemy target. This general range was then used by the artillery to determine the load and trajectory. Corrections were made by trial and error. As artillery has become more accurate, the accuracy and speed of calculations needed to be improved. Today computers calculate distances using the functions that define the triangulation method used in this activity.

Management

1. Make the transits prior to doing this activity. They can be made by either the teacher or the students. Follow the instructions on *Target Practice: Making the Transit.*
2. The transits are used most effectively by two students. Have one student hold the transit and align it with the baseline by pointing the arrow towards another landmark on the baseline. The second student will sight through the straw that has been taped to the transit to make sure the transit is accurately aligned. While the first student maintains the aligned position by continuing to sight through the straw at the landmark on the baseline, the second student sights the unknown landmark through the movable straw. When the second student has carefully sighted through the movable straw, the angle of rotation can be read on the side of the transit closest to the unknown object.
3. To provide the class with practice for using this method of measurement, the activity sheets present the problem of shooting a ship's cannon at a target on land (see *Key Question*) which helps students visualize the process.
4. Before taking a class outside to do this activity, the teacher should establish a baseline. This line should be easily recognizable and stretch most of the way across the area. The edge of the surfaced playground, sidewalk, or marked boundary of a sports field make good baselines. If such a line is not available, stake down a long string to serve as a baseline.
5. Identify two points with some markers (flags or sticks) along the baseline. One marker will represent the bow of the ship, the other the stern. The markers should be separated by a distance that represents the length of the ship. The student drawing page has the bow and stern separated by 15 cm. By placing the markers 15 meters apart, the scale drawing will be simpler for the students. With these dimensions the scale will be 1 cm= 1 m. The story line has one sighter in the bow near the cannon, and another in the stern.
6. The teacher will also need to identify an object that is well away from the baseline to serve as the target. The landmark should be a permanent object such as a tree, sign, post, or building.
7. Prior to this activity, the teacher may choose to measure the distance between the marker serving as the bow and the landmark serving as the target. This will allow students to check their accuracy.

Procedure

1. Discuss the *Key Question.*
2. Distribute transits and demonstrate how they are used.
3. Distribute the student sheet and discuss the procedure with students.
4. Take the class out to the area with the baseline to take and record the necessary measurements.
5. Have the students return to the classroom and make scale drawings using the *Student Drawing Sheets* and following instructions. If the landmarks on the baseline were not separated by 15 meters, help them determine an appropriate scale.
6. Guide students to measure the scale drawing and use proportions to determine the distance to the target.
7. Direct them to compare their calculated distances to the actual distance to check accuracy.
8. Discuss outcomes and improvements.

Discussion

1. How accurate was your calculated distance?
2. How could you improve your accuracy? [make more careful measurements; improve drawing techniques; extend the distance between sighters on the baseline]
3. What are some other applications for this method of measurement? [surveying, navigation]

Extensions

1. Allow students to practice this measurement method. Place matching colored or labeled flags on bamboo skewers. Place one flag out in the field to serve as the benchmark and a pair along the baseline area where sightings are to be made. As an assessment, have students use indirect measurement to determine the distances between the two sets of flags.
2. Set up a treasure hunt where students are given two landmarks that establish a baseline and the angle of sight to the treasure from both landmarks. Have students draw a treasure map and determine the location of the treasure.

Target Practice

Making the Transit

1. Cut out transit and mount on cardboard.
2. Tape one straw along the line that runs through the middle of the arrow.
3. Poke a straight pin through the middle of another straw. Push the head until it is flush with one side of the straw and the other end of the pin sticks out the opposite side.
4. Poke the end of this pin through the straw that is taped to the transit so that the end of the pin goes through the cross hairs in the small circle on the transit. The top straw is now like the pointer on a spinner.
5. Push the pin so that it sticks out on the back side of the cardboard, but does not crush the straws. Bend the end of the pin so it is flat on the backside of the cardboard. Tape the end of the pin down so no one will get poked.

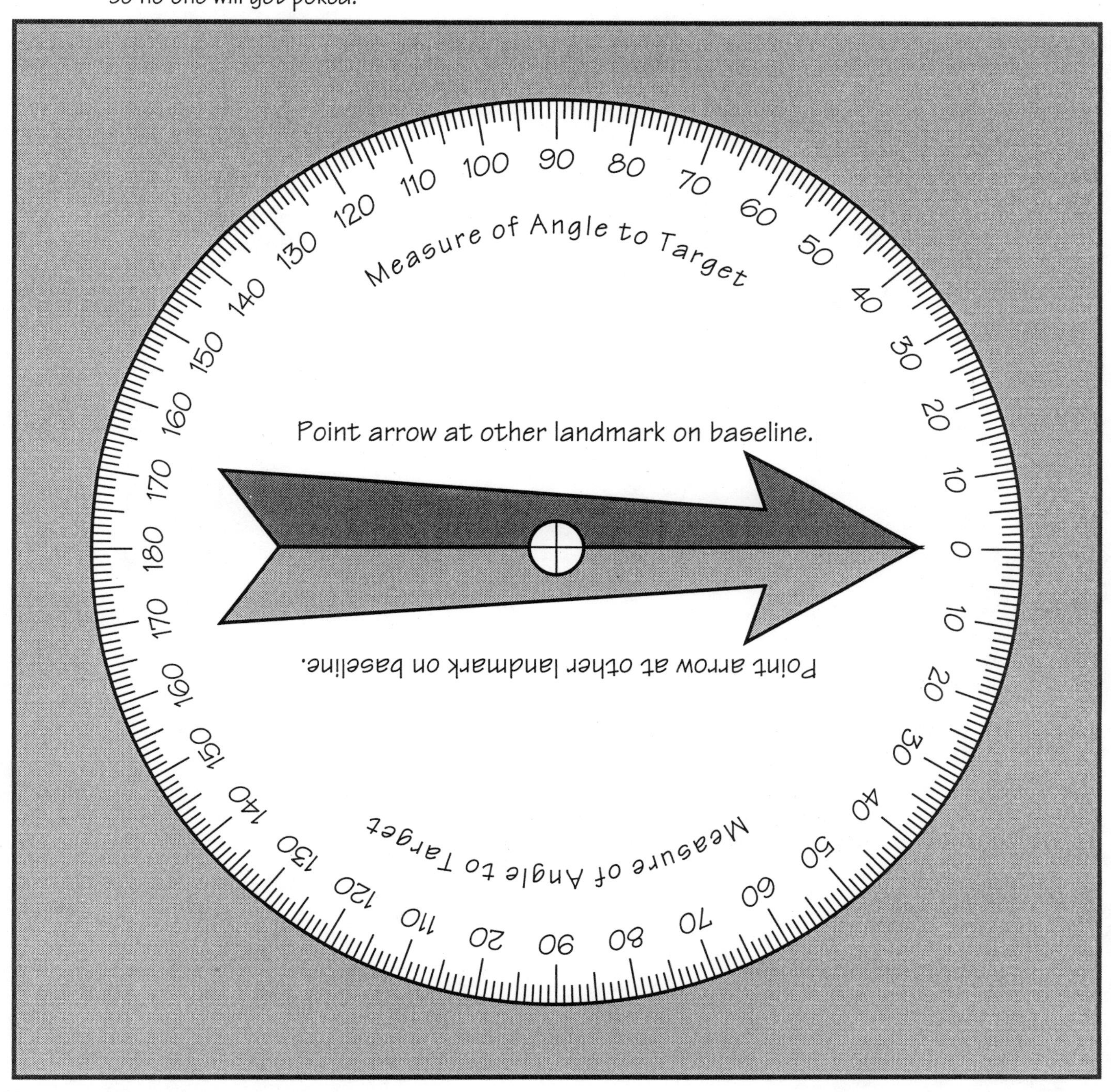

Target Practice

If you were going to fire a cannon from a ship, how could you determine how far away the target was?

Take outdoor measurements

1. Measure the length of the baseline from the sighter in the bow to the sighter in the stern. Record the distance on the drawing.
2. The sighter at the bow measures the angle between the baseline of the ship and the target. Record this angle measurement in the space on the drawing.
3. The sighter at the stern measures the angle between the baseline of the ship and the target. Record the angle measurement in the space on the drawing.

Make a scale drawing

4. Draw a line on the paper to represent the baseline. Scale the length to represent the distance from the bow to the stern. (1 cm = ______ m)
5. Place a protractor at the end of the line that represents the bow of the ship and measure an angle equal to the one in the sighting. Extend the ray that represents the sighting.
6. Place a protractor at the end of the line that represents the stern of the ship and measure an angle equal to the one in the sighting. Extend the ray that represents the sighting.

Calculate the distance to the target

7. Measure the sighting line from the bow to where it intersects the other sighting line. Record the length in the proportion.
8. Record the scale in the proportion.
9. Perform the necessary calculation to complete the proportion and determine the distance to the target.

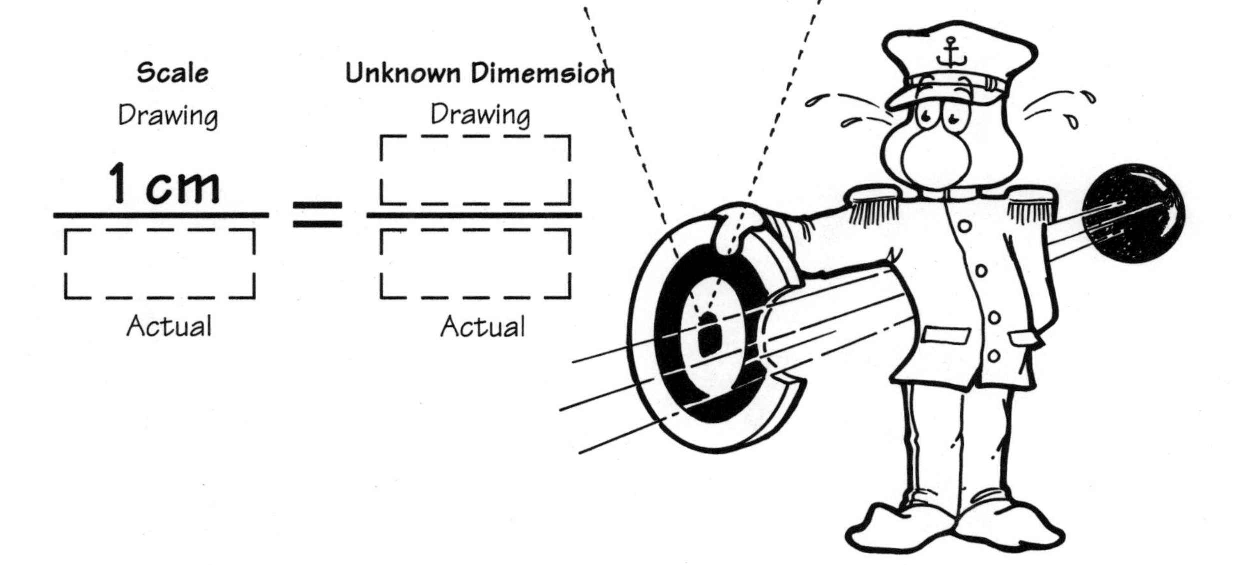

Topic
Making scale maps

Key Questions
1. How did early explorers make accurate maps?
2. How could you make a map of the school building quickly and accurately?

Focus
Students will make a map using a plane table. The plane table method allows the similar triangles to be drawn directly to scale on the paper.

Guiding Documents
NCTM Standards
- *Verify and interpret results with respect to the original problem situation*
- *Reflect on and clarify their own thinking about mathematical ideas and situations*
- *Validate their own thinking*
- *Understand and apply geometric properties and relationships*

Project 2061 Benchmarks
- *It takes two numbers to locate a point on a map or any other flat surface. The numbers may be two perpendicular distances from a point, or an angle and a distance from a point.*
- *Estimate distances and travel times from maps and the actual size of objects from scale drawings.*

Math
Measurement
  linear
  angular
Geometry
Proportional reasoning
Scale drawings

Social Science
Geography
  map making
History
  exploration

Integrated Processes
Observing
Measuring
Collecting and recording data
Interpreting data
Applying

Materials
Per group:
  clipboard or 1" x 12" x12" board
  large can
  ruler
  tape measure or measuring rope
  paper
  masking tape
  alidade or triangular scale rule or ruler and small nails (see *Management 4*)

Background Information
The plane table method requires three pieces of equipment:
a. The plane table, a drawing board that can be leveled. Historically, this was a board attached to a tripod.
b. An alidade, a straightedge with a sight running parallel to the edge.
c. A tape measure or measuring instrument to determine the distance between sightings.

The plane table method of map making was used by explorers to make maps. Sightings could be made at several prominent locations to other significant landmarks. With these few points, the map makers would fill in the other details by sight. Explorers could even make sightings from the deck of a ship if the seas were calm. The plane table method of mapping allowed an explorer to make a map while moving through an area. The method of using the plane table is explained in the student instruction sheets.

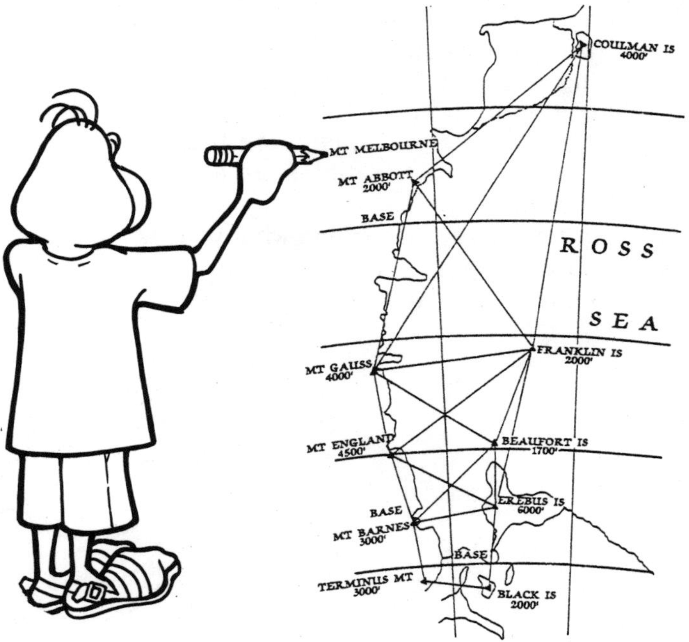

Management

1. In this activity a board or clipboard and a large can will serve as the plane table. The board may be any piece of lumber that is as large as an 8 1/2" x 11" piece of paper. Large cans from the cafeteria provide an easily accessible platform for the table.
2. Students will need to make the table as level as possible when making sightings. This is accomplished by filling the can with water up to one of its rings. When taking sightings, students can check if the can is level by seeing that the water is at the same ring all around the can. Level adjustments are made by placing objects under the can or modifying the soil in the area.
3. Students will need to lie on the ground to do the sightings unless they raise the plane table. They can raise the height of the plane table by placing it on a tall stool or an inverted wastebasket.
4. A triangular scale rule works well as an alidade. If such is not available, an alidade can easily be made by placing two nails in a ruler. The nails should be at opposite ends of the ruler, several centimeters from the ends. Both nails need to be the same distance from the straight edge of the ruler. By using the nails (or top edge of the triangular rule) as a sight and putting them in line with a landmark, the alidade is aligned with your line of sight. A line drawn along the straight edge is parallel to the line of sight and is used to draw lines on the map that represent the line of sight.
5. The procedure for this activity is written explaining how students should map a school building. Before doing this, it is suggested that the teacher first model the technique and then have the students practice on sighting one landmark.
6. Set up the area similar to *Target Practice*. Students can determine distances between the sighting points and the landmark by measuring the sides of the triangle that is drawn. Have students note the similarity between the plane table method and using a transit. This initial session works well inside a large room such as a cafeteria or gym where tables help to guarantee that sightings are level and the plane tables are at a more reasonable height.
7. Before beginning the activity, determine what you want the students to map. A school building as shown in the example may be used. Other suggestions include a playing field or baseball backstop using several landmarks.
8. All mapping has an amount of error and depends on the precision of the instruments and the skill of the surveyor. Considering the crude instruments used it this activity, some error will occur. Students should discuss how they can identify error and possible ways they can keep it to a minimum. Having students find their percent of error will help them to make quantitative judgments and comparisons of their methods and skills.

Procedure

1. Discuss the *Key Questions.*
2. Distribute materials.
3. Have students read through instructions and make sure they understand the procedure.
4. Following the instructions, guide students in mapping the area.
5. Discuss accuracy and have students check their maps.

Discussion

1. How accurate is your map? (Have students calculate distances using the map, and then have them check the accuracy outside.)
2. How could you improve the accuracy of your map if you were to do it again? [be more careful in procedure; lengthen distance between sighting points; use more defining points]

Extensions

1. Make a map from the center of a field so that the baseline is down the middle, with defining landmarks in all directions.
2. Look at copies of explorers' maps and compare them to modern maps. Compare the accuracy and detail.
3. Make a map of a larger area as described in *Out-Landish Mapping.*

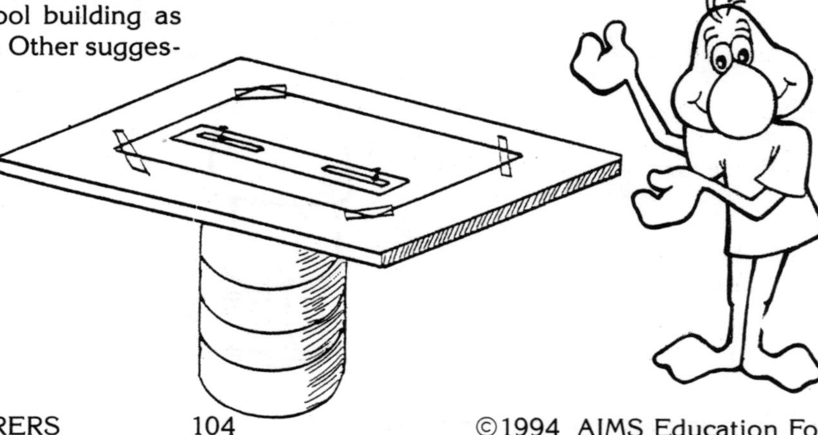

Instruction Sheet

1. Tape a sheet of blank paper to the board.
2. Fill the can with water so it is level with one of the rings, and go to the area you plan to map.
3. Look at what you are to map and determine what points could be used to define the area. (A building is defined by its corners. Open areas may be defined by fence corners, trees, posts, structures, large rocks and other objects. Markers may be placed if natural landmarks do not occur.)

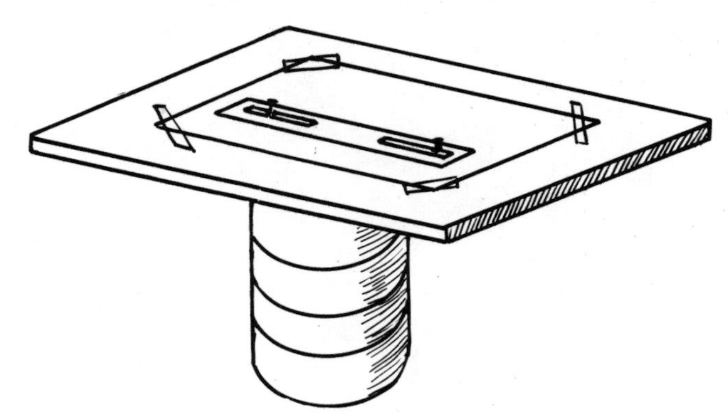

4. Decide where to make the baseline for your map. It should be to one side of all the defining landmarks and not in line with any of them.
5. Use either markers or landmarks to identify two points on the baseline.
6. Use the straight edge of the alidade to draw a line near the bottom of the paper on the board. This line represents the baseline.
7. Place the plane table (can with board on top) at one of the points on the baseline. Check that the board is level by seeing that the water is at the same ring all around the can. If it is not level, put objects under edges of the can to level the water.
8. On the map, draw a point on the baseline and label it "A." This represents your present location.
9. Place the straight edge of the alidade along the baseline of the drawing.
10. Rotate the drawing board without moving the alidade or disturbing the level of the can until the sight of the alidade lines up with the other landmark on the baseline . The baseline of the map is now in line with the actual baseline.
11. Without moving the plane table, pick up the alidade. Put the point of your pencil on the point on the drawing that represents your location. Put the edge of the alidade next to the pencil. Rotate the alidade around the pencil until it lines up with one of the defining landmarks.
12. Draw a line along the edge of the alidade to represent the line of sight to that landmark.

13. Draw a line to each defining landmark following the procedure of the previous two steps. Your map should look similar to the one above.

14. Measure the distance to the second landmark on the baseline.
15. Move the plane table to the second landmark and level it again.
16. Decide on an appropriate scale such as 1 cm = 2 m. Determine how far in scale the second point on the baseline should be from the first point.
17. Measuring with the ruler, place a second point on the map's baseline that is the scaled distance from the first point. Label the second point "B."
18. Follow steps 9 through 13. Your map should look similar to the one on the right.

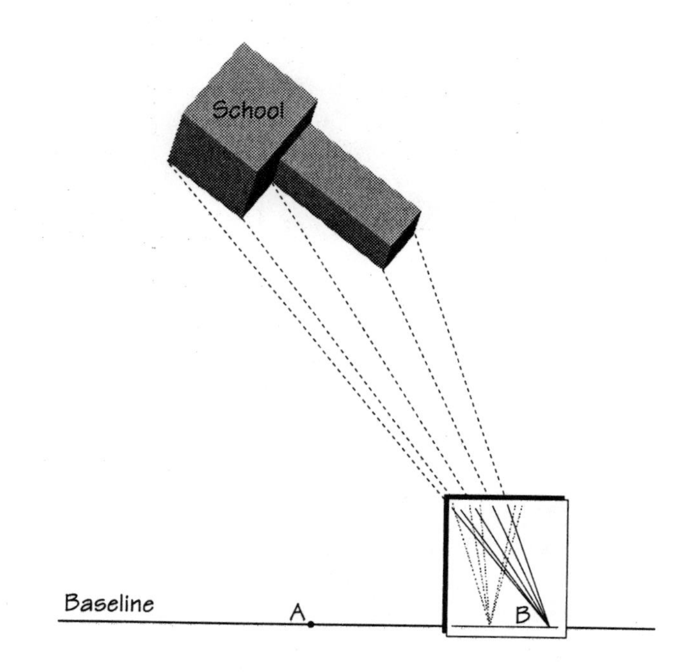

19. The intersections of the corresponding lines of sight on the map represents the defining landmarks. The whole map is drawn to the scale you chose in step 16.
20. In the case of a building, connecting the defining points gives you the outline of the building.

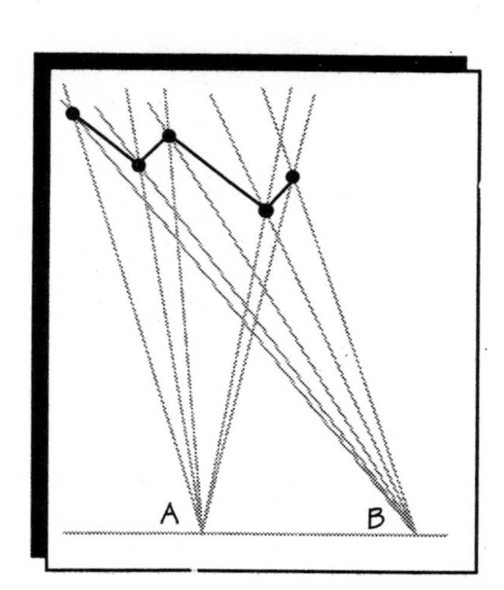

Out-landish MAPPING

Topic
Making scale maps

Key Question
How could you make an accurate map of a large area?

Focus
Students map a larger area using the plane table method.

Guiding Documents
NCTM Standards
- *Develop an appreciation of geometry as a means of describing the physical world*
- *Use mathematics as an integrated whole*
- *Extend their understanding of the concepts of perimeter, area, volume, angle measure, capacity, and weight and mass*
- *Develop formulas and procedures for determining measures to solve problems*

Project 2061 Benchmarks
- *It takes two numbers to locate a point on a map or any other flat surface. The numbers may be perpendicular distances from a point, or an angle and a distance from a point.*
- *Estimate distances and travel times from maps and the actual size of objects from scale drawings.*

Math
Measurement
 linear
 angular
Geometry
Proportional reasoning
 scale drawings

Social Science
Geography
 map making

Integrated Processes
Observing
Collecting and recording data
Applying

Materials
For student groups:
 plane tables (see *Plane Table Mapping*)
 alidades (see *Plane Table Mapping*)
 tape measures
 paper
 tape
 colored paper
 bamboo skewers
For the class:
 Vellum, 24" x 36" or 30" x 42" (see *Management 5*)
 protractor
 magnetic compass
 transit (see *Target Practice*)
 black fine-tip marker

Management
1. Before doing this activity find a suitable area to map. Small lakes or ponds are excellent areas. Open areas with a winding creek, river, or road running through them also work well. City parks often provide such an environment.
2. If interesting geographical forms are not easily accessible, a large irregular area may be outlined on a playing field.
3. An extended period of time is needed to complete the map. You may choose to make this an all-day outing. If the location is close to school, the mapping may be done the same period each day for a week.
4. All mapping has an amount of error depending on the precision of the instruments and the skill of the surveyor. Considering the crude instruments used it this activity, some error will occur. Students should discuss how they can identify error and how they can keep it to a minimum. This will become apparent as the student zones are matched and traced to the sighted zone separation markers.
5. Vellum is large tracing paper for drafting and is available at stationery stores or art supply stores.

Site preparation prior to the activity:

6. Go to the site before taking the class. (The preparation can be used as a whole class activity if it is appropriate for the class.)
7. Visually survey the area to determine how the area could be divided into zones for mapping. Each group of students will survey a zone.
8. Identify landmarks that separate the zones.
9. Get a rough idea of the area's measurements either by pacing or direct measurement.
10. Determine a scale that will allow the map to fit on the sheet of vellum and each zone to fit on the students' sheets.

11. Identify two landmarks that establish a baseline that can be measured accurately. The landmarks do not have to be the ones used for zone separations.
12. Measure the distance between the two landmarks on the baseline.
13. With the transit, stand at each landmark on the baseline and measure the angle between the other landmark on the base line and every zone separation landmark. Record all the measurements.
14. Stand at one of the landmarks on the baseline and use a compass to measure the angle between geographic north and the baseline (see *Treasure Hunt*). Record the measurement.
15. Return to the classroom and draw the baseline to scale on the vellum.
16 Draw in the sighting lines from each point representing the baseline landmarks. The intersections of corresponding sight lines represent the zone separation landmarks. (Note: You may want to borrow a commercial transit as these measurements establish the accuracy of the map.)

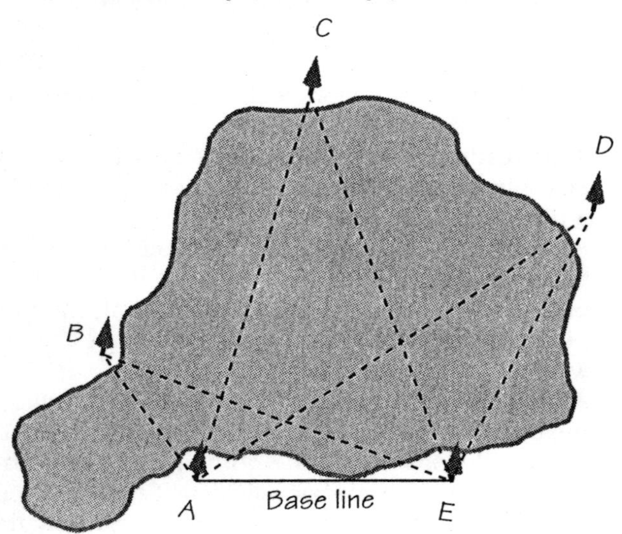

Figure 1: The area to be surveyed is a small lake. The pine trees divide the shoreline into five zones. Landmarks A and E are the ends of the baseline. The dotted lines represent the sight lines.

Procedure

Student Mapping:

1. Before leaving for the site, make sure students are familiar with the procedures. Students should have completed *Plane Table Mapping*.
2. On arrival at the site, identify the zone separation landmarks for the students. Assign a zone to each group.
3. Each group visually surveys their zone and identifies points that could be used to roughly define the zone. If these points do not have landmarks, markers need to be placed. The markers can be made from colored paper and bamboo skewers.

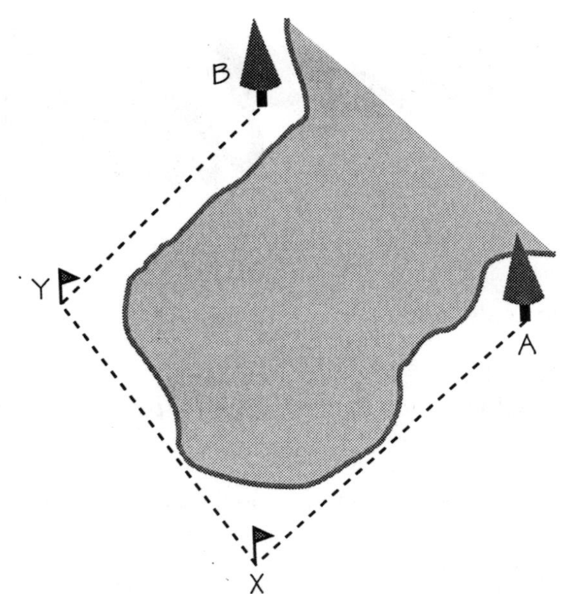

Figure 2: This group has been assigned the zone between *Landmarks A* and *B*. *Markers X* and *Y* help to roughly define the zone.

4. Students do a plane table map between the first two landmarks in the zone, sighting to each of the other landmarks. The scale used on the line is the same as the one used on the baseline on the class map done on the vellum.
5. The students continue the map with lines between the second and third landmarks, the third and fourth, and so forth until they have sighted from all the landmarks. Some of the sight lines become the rough outline of the zone.
6. As students are mapping, they should be checking for accuracy. If the intersections of three corresponding lines do not meet at one point, there is an inaccuracy in at least one of the sightings and they should be encouraged to check the sightings again to make corrections.

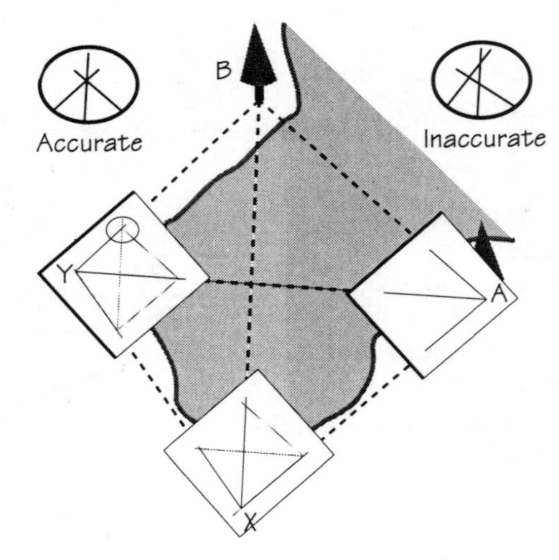

Figure 3: The rough map of the zone is made. Plane table sightings are taken from each landmark to every other landmark. The plane table illustration at *Point X*

shows how to make a change in directions. The magnified ovals at the top are made from the table at *Landmark Y*. They show the difference between and accurate and inaccurate mapping.

7. More accurate positioning of land forms and landmarks is done by making measurements that are perpendicular to the lines between defining landmarks. In the case of a lake, river, or road, this can be done by starting at the beginning landmark and making perpendicular measures at regular intervals. (Students may use the transit or sight along the edge of a sheet of paper to measure the right angle. Another method of determining a 90° angle is the Wing Ding method. To do this, stand along the line on which you want to find the perpendicular. Stretch your arms out to the sides, close your eyes, and bring your hands together in front of you as illustrated below. It is very important to close your eyes so you will not unconsciously be drawn from the 90° to an existing landmark.) The distance is scaled and the perpendicular line is drawn on the map.

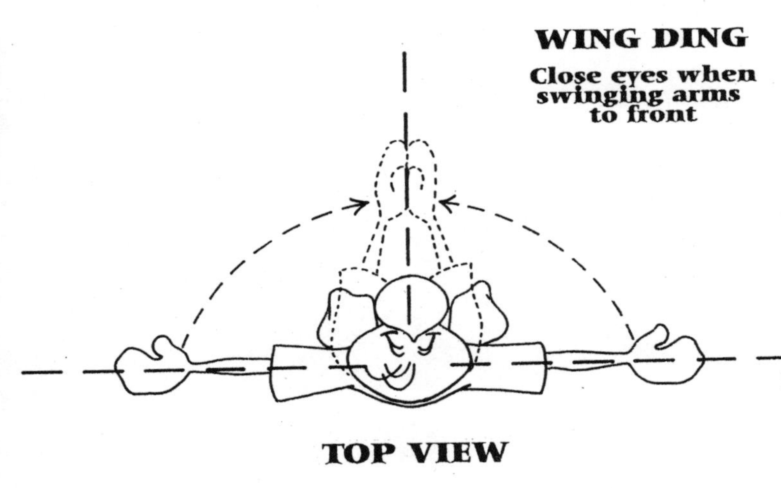

8. The final lines and details are made visually. In the case of a lake, river, or road, the ends of the perpendicular lines are connected while observing to the area. Other details can be sketched in using measurements from established points.

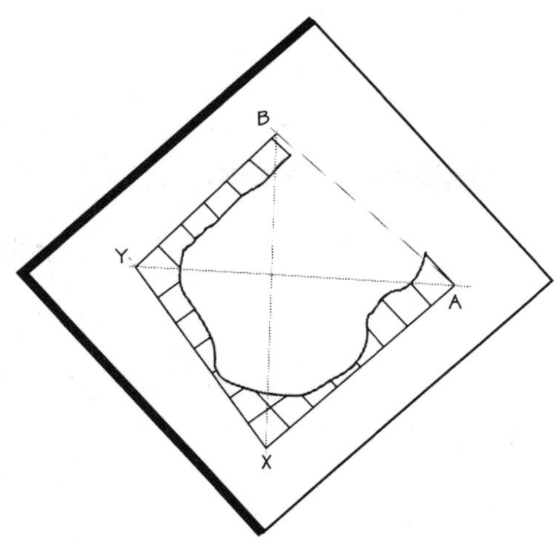

Figure 4: The rough outline is detailed by drawing in the perpendicular scaled lines at regular intervals. The final outline of the lake is drawn by connecting the ends of the lines with visual confirmation.

Completion of map

1. Place each student map under the vellum so that the ends of the rough outline are directly under the corresponding zone separation landmarks. If the points do not correlate exactly, either the student map, or the landmarks on the vellum is inaccurate and reconciliation will need to be made at the teacher's discretion
2. Trace the final outline of the map onto the vellum. Do not trace sight lines or the rough outline.
3. Place the title, legend (a key or brief description), scale, and compass rose (an illustration indicating directions) on the map.
4. Ink in the final map using black fine-tip markers.
5. Have the map reproduced as blue line drawings and distribute to the students.

Discussion

1. Why does everyone need use the same scale? [Each zone map is traced onto the same master.]
2. How could you tell if you were making accurate sightings and measurements. [Sighting intersections should meet. Actual measurements don't match scaled measurements. Zone maps do not match sighted zone separation landmarks on vellum.]
3. How could you eliminate these errors in your measurements? [Check level of sighting surface. Remeasure angles and lengths in question.]

SPACE MAPS

Topic
Topographical mapping and space exploration

Key Question
How can digital data be made into a picture showing us what a planet looks like?

Focus
Students will determine the pattern used to record digital data and how that data represents a geographical form. From the data they will construct a model and draw a topographical map and a perspective drawing.

Guiding Documents
NCTM Standards
- *Model situations using oral, written, concrete, pictorial, graphical, and algebraic methods*
- *Understand and apply reasoning processes, with special attention to spatial reasoning and reasoning with proportions and graphs*
- *Make inferences and convincing arguments that are based on data analysis*

Project 2061 Benchmarks
- *Technology is essential to science for such purposes as access to outer space and other remote locations, sample collection and treatment, measurement, data collection and storage, computation, and communication of information.*
- *Information can be carried by many media, including sound, light, and objects. In this century, the ability to code information as electric currents in wires, electromagnetic waves in space, and light in glass fibers has made communication millions of times faster than is possible by mail or sound.*
- *Most computers use digital codes containing only two symbols, 0 and 1, to perform all operations. Continuous signals (analog) must be transformed into digital codes before they can be processed by a computer.*

Math
Patterns
Scale drawing
Graphing
Visualization skills
Topographical mapping
Dimensions

Science
Earth science
topographical mapping
space exploration

Integrated Processes
Observing
Comparing and contrasting
Analyzing data
Interpreting data
Modeling

Materials
700 interlocking 2-cm cubes
Colored pencils
Straight edges

Background Information
Exploring and mapping Earth has been relatively simple since we live on this planet and are able to take measurements first hand. As we have ventured out into space, mapping has become more difficult. We now depend on machines, satellites and space probes, to take measurements and gather data for us. To further complicate the situation, this information has to be sent millions of miles.

Engineers and scientist choose to have all information sent to and from the space craft in the form of numbers. Even pictures are translated into numbers because numbers are easy for the computers on the space craft to deal with. These numbers are called digital data.

Space probes send all their data back to earth as digital data. We do not receive pictures of planets; we receive billions of numbers which describe various dimensions of the planet at numerous locations.

Any single location on the planet will have many numbers recording its dimensions. To determine the location's position, the three dimensions of longitude, latitude, and elevation are required. (These are the three used in this activity. Other data might include temperature, chemical make-up, or reflectivity.)

Algorithms are used both to compile the data for transmission and for translation after reception. These algorithms organize the data so it can be understood and analyzed. These organizational systems were developed by electrical engineers and computer scientists. Like all codes, they have a pattern that allows them to be deciphered by others; however, without understanding the code, the data are useless numbers.

Scientist also have found that when information is sent through space as digital data, it has less errors. What happens to the picture on your television in a thunderstorm? You get static and it messes up the picture. Television pictures are not sent as digital data so they are easily distorted and have errors. The same thing would happen in space so pictures are translated into numbers by computers, and the numbers are sent. Digital data allows space craft to send a lot of information, very rapidly, over great stretches of space, with few errors.

Digital data transmissions were chosen over analog transmissions (i.e. commercial television broadcast) for reasons which include:

- Digital data are much more flexible in the types of information that can be sent.
- Digital data are more accurate.
- Digital data are much easier to analyze and manipulate with a computer.

The data in this activity are simulated to allow students the opportunity to see how digital data can be organized to provide a geographical understanding. The first two activity sheets deal with the same imaginary mountain. The third sheet represents the crater of Mount St. Helens.

Management

1. Before doing this activity students should have an understanding of what topographical maps are and how they are made.
2. This activity has several levels of completion. The teacher must decide what is appropriate for the students and exit at that level.
   a. The students may just discern the pattern and make a topographical map.
   b. Students may continue by making a model of the map.
   c They may complete the whole activity by making a perspective drawing from the data.
3. See *Extension* for ordering an applicable video from NASA Jet Propulsion Laboratory.

Procedure

Making the Map

1. Distribute the *Data Output One* and have students discuss the *Key Question*.
2. Allow students plenty of time to discover patterns in the data. Have them share their discoveries as they make them. (Students should observe that the numbers are arranged in blocks of three. The first or top number in each block is always the same in each row of blocks. This first number gets greater as you go down the column. The top number represents the latitude. The second number is the same in each column, and gets greater as you go across any row. It represents the longitude. The bottom number has no discernible pattern. It represents the location's elevation.)
3. Have students construct a topographical map on *Data Output One* by drawing a curve enclosing the positions of highest elevation. Direct them to draw a larger concentric curve around the positions with the next lower interval of elevation (intervals are 20 units). Have them continue drawing concentric curves until the map is complete.
4. Ask students to provide a physical description of the land form represented by the map. (It's a mountain.)

Making the Model

5. Distribute *Build a Model* and the 2-cm cubes to each group.
6. Assign each group a row to build by stacking cubes for each square. Each cube represents 20 units of elevation. The stacks are then placed next to each other in order to make a cross-section of the mountain.
7. Have each group bring up their cross-section of the model and place it on the proper row on a *Build a Model* page which has been placed on a table. When all the cross-sections are assembled, there will be a model of the mountain.
8. Have students make a relief map on their model sheets by coloring in each altitude range the same color.
9. Using the color coding, guide students in making a contour map on top of the relief map.

Making the Drawing

10. Distribute the *Drawing: Data Set One.* Explain to the students that they will be making a drawing of the model. The drawing will be done as if they are viewing the model from the lower right corner (southeast).
11. The numbers to the left label the column data (middle number in set) and the numbers on the right label the rows (first number in set). The grid shown is at an elevation of 0. The vertical space between dots represents 20 units of elevation.
12. Have students start with the bottom row of data (row 1100). Have them put a mark with a pencil on the dot above each column that is the scaled height at that position. For column 200, row 1100, the third dot up is marked because an elevation of 60 units equals 3 dots.
13. When the altitudes of the ten positions in the row have been marked, connect them to the adjacent marks in that row as shown in the figure below.

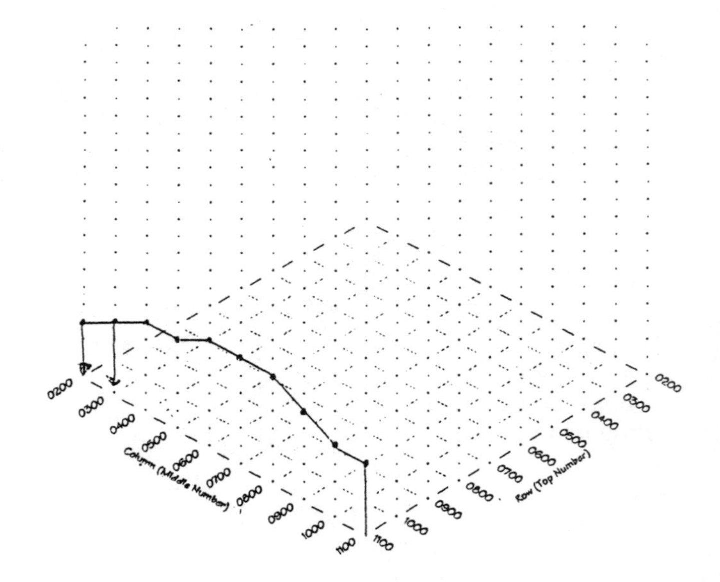

14. Direct the students to mark the altitudes of the 10 positions in row 1000. They must be aware that these altitudes are placed directly above the bold points at the intersections of the rows and columns.
15. Have them connect the marked altitudes in row 1000 to the adjacent altitudes in that row. They should then draw a line between each altitude in row 1000 to the corresponding altitude in the preceding row as shown in the figure below.

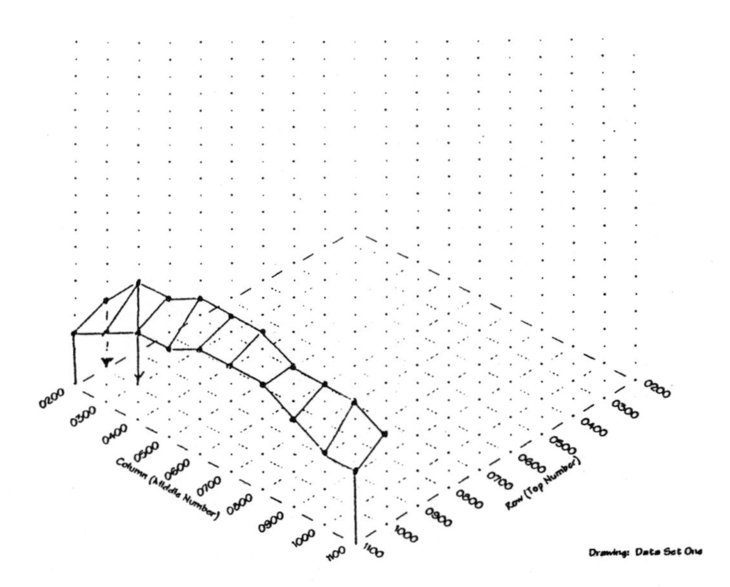

16. Have students complete each row by following the procedure outlined in steps 14 and 15.
17. When students have penciled in the drawing, they may choose to ink in the lines with a marker to see the drawing more clearly.
18. If desired, students may complete a map and model on their own with *Data Set Two*.

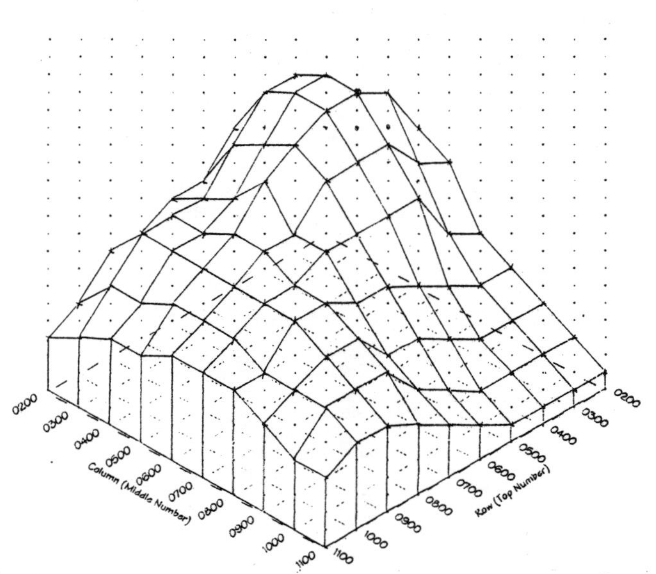

Drawing: Data Set One

Discussion

1. How many units of distance does each edge of a cube represent? [100]
2. How many units of elevation does each cube represent? [20]
3. How does our model look different than the actual mountain? [Elevation is distorted 5 times. The real mountain would be appear much shorter.]
4. How is the drawing distorted? [vertical distortion: 100/40 = 2.5; the mountain is drawn 2.5 times higher than it is.]
5. What other types of dimensions might a space probe send back about a planet? [temperature, chemical make-up, reflectivity]

Extension

NASA has a video that has some demonstrations of how computers have made perspective drawings. The video has been animated so it appears that you are flying over Earth and part of Mars. It helps students see the application of what they are studying, besides providing high motivation and appreciation for the activity. The video is available by sending a sealed, blank, high-quality video cassette (T-120 VHS format) to the address listed below. Request that they make you a copy of the video *JPL Computer Graphics*. The selections related to this activity include: *L.A. - The Movie; Flight to the Moon: Mars - The Movie; Miranda - The Movie.*

Teacher Resource Center
NASA Jet Propulsion Laboratory
Mail Code CS-530
4800 Oak Grove Drive
Pasadena, CA 91109

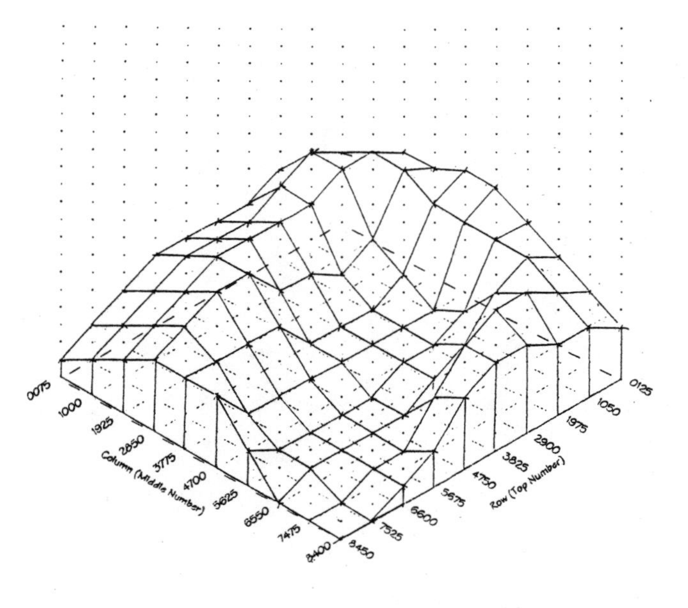

Drawing: Data Set Two

Data Output One

Below is a data print-out of a satellite transmission from an exploratory probe of a planet. The satellite was on a mapping mission. The satellite's relay was malfunctioning so this is the only data received. Your job is to decipher what information is being given by data and produce a means of displaying your findings.

0200	0200	0200	0200	0200	0200	0200	0200	0200	0200
0200	0300	0400	0500	0600	0700	0800	0900	1000	1100
0120	0140	0160	0180	0160	0100	0080	0060	0040	0020
0300	0300	0300	0300	0300	0300	0300	0300	0300	0300
0200	0300	0400	0500	0600	0700	0800	0900	1000	1100
0140	0180	0220	0240	0180	0120	0080	0060	0040	0020
0400	0400	0400	0400	0400	0400	0400	0400	0400	0400
0200	0300	0400	0500	0600	0700	0800	0900	1000	1100
0160	0240	0260	0260	0220	0180	0100	0060	0040	0020
0500	0500	0500	0500	0500	0500	0500	0500	0500	0500
0200	0300	0400	0500	0600	0700	0800	0900	1000	1100
0180	0240	0260	0260	0220	0180	0120	0080	0040	0020
0600	0600	0600	0600	0600	0600	0600	0600	0600	0600
0200	0300	0400	0500	0600	0700	0800	0900	1000	1100
0140	0200	0220	0240	0220	0180	0140	0100	0060	0040
0700	0700	0700	0700	0700	0700	0700	0700	0700	0700
0200	0300	0400	0500	0600	0700	0800	0900	1000	1100
0140	0160	0180	0220	0180	0180	0140	0100	0080	0060
0800	0800	0800	0800	0800	0800	0800	0800	0800	0800
0200	0300	0400	0500	0600	0700	0800	0900	1000	1100
0120	0160	0160	0180	0180	0160	0160	0140	0120	0080
0900	0900	0900	0900	0900	0900	0900	0900	0900	0900
0200	0300	0400	0500	0600	0700	0800	0900	1000	1100
0120	0160	0160	0160	0160	0160	0160	0140	0120	0100
1000	1000	1000	1000	1000	1000	1000	1000	1000	1000
0200	0300	0400	0500	0600	0700	0800	0900	1000	1100
0080	0120	0120	0140	0140	0140	0120	0120	0120	0100
1100	1100	1100	1100	1100	1100	1100	1100	1100	1100
0200	0300	0400	0500	0600	0700	0800	0900	1000	1100
0060	0080	0100	0100	0120	0120	0120	0100	0080	0080

Are there any patterns in the data that might give you a clue to what this information is conveying?

Build A Model

Use interlocking cubes to build a model of this planet's features. Make columns with one cube representing each contour interval of altitude. Then color in each square of the same altitude range one color. Finally, draw in the contour lines.

0200 0200 0120	0200 0300 0140	0200 0400 0160	0200 0500 0180	0200 0600 0160	0200 0700 0100	0200 0800 0080	0200 0900 0060	0200 1000 0040	0200 1100 0020
0300 0200 0140	0300 0300 0180	0300 0400 0220	0300 0500 0240	0300 0600 0180	0300 0700 0120	0300 0800 0080	0300 0900 0060	0300 1000 0040	0300 1100 0020
0400 0200 0160	0400 0300 0240	0400 0400 0260	0400 0500 0260	0400 0600 0220	0400 0700 0180	0400 0800 0100	0400 0900 0060	0400 1000 0040	0400 1100 0020
0500 0200 0180	0500 0300 0240	0500 0400 0260	0500 0500 0260	0500 0600 0220	0500 0700 0180	0500 0800 0120	0500 0900 0080	0500 1000 0040	0500 1100 0020
0600 0200 0140	0600 0300 0200	0600 0400 0220	0600 0500 0240	0600 0600 0220	0600 0700 0180	0600 0800 0140	0600 0900 0100	0600 1000 0060	0600 1100 0040
0700 0200 0140	0700 0300 0160	0700 0400 0180	0700 0500 0220	0700 0600 0180	0700 0700 0180	0700 0800 0140	0700 0900 0100	0700 1000 0080	0700 1100 0060
0800 0200 0120	0800 0300 0160	0800 0400 0160	0800 0500 0180	0800 0600 0180	0800 0700 0160	0800 0800 0160	0800 0900 0140	0800 1000 0120	0800 1100 0080
0900 0200 0120	0900 0300 0160	0900 0400 0160	0900 0500 0160	0900 0600 0160	0900 0700 0160	0900 0800 0160	0900 0900 0140	0900 1000 0120	0900 1100 0100
1000 0200 0080	1000 0300 0120	1000 0400 0120	1000 0500 0140	1000 0600 0140	1000 0700 0140	1000 0800 0120	1000 0900 0120	1000 1000 0120	1000 1100 0100
1100 0200 0060	1100 0300 0080	1100 0400 0100	1100 0500 0100	1100 0600 0120	1100 0700 0120	1100 0800 0120	1100 0900 0100	1100 1000 0080	1100 1100 0080

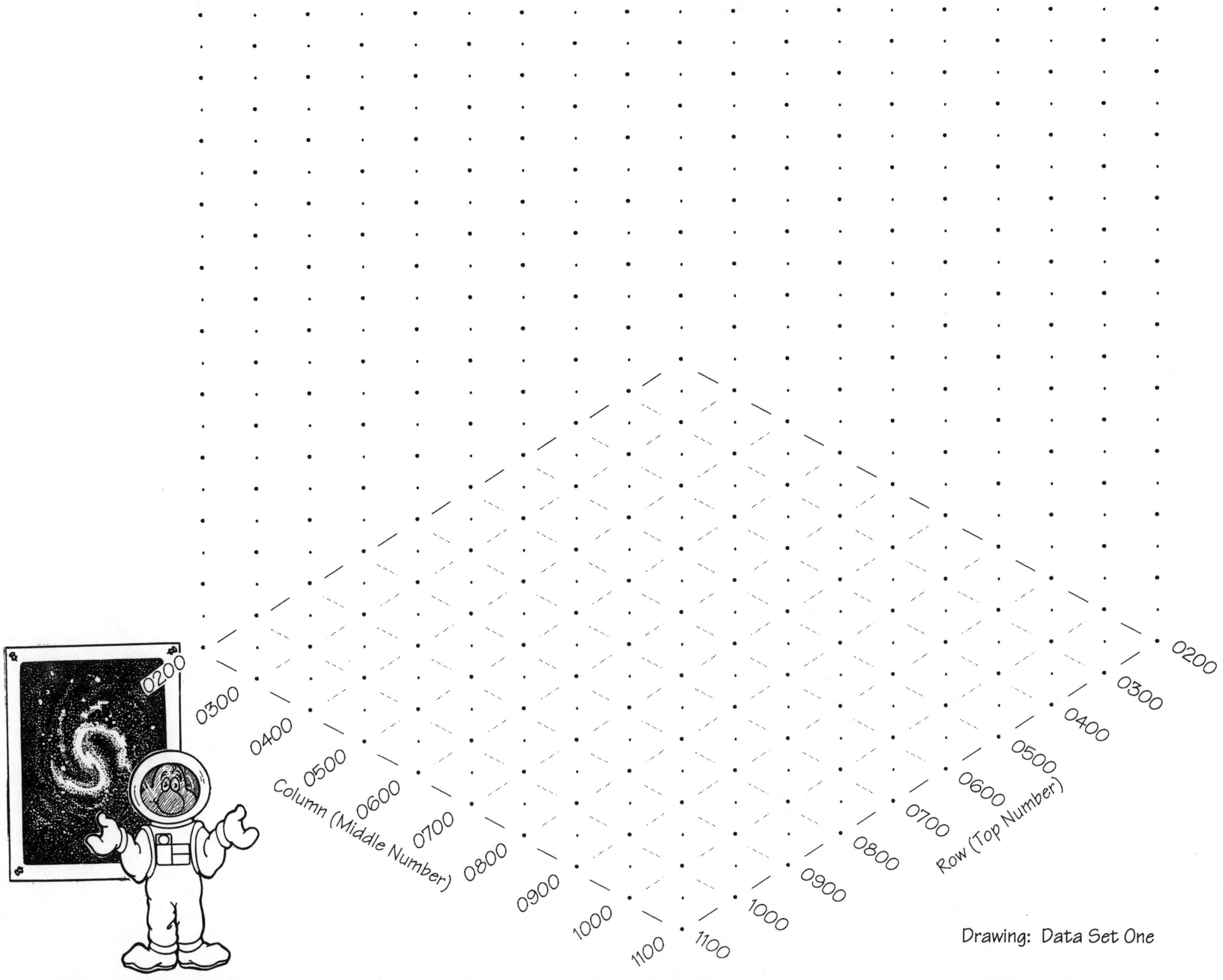
0200
0300
0400
0500
0600
0700
0800
0900
1000
1100
Column (Middle Number)
0200
0300
0400
0500
0600
0700
0800
0900
1000
1100
Row (Top Number)
Drawing: Data Set One

Data Output Two

Use the data below from an earth-orbiting satellite to make a three dimensional representation of the landform. Horizontal sightings were taken at 925-foot intervals. Vertical readings were rounded to the nearest 400-foot interval.

0125	0125	0125	0125	0125	0125	0125	0125	0125	0125
0075	1000	1925	2850	3775	4700	5625	6550	7475	8400
6800	7200	7600	7600	8000	8000	7600	7200	6800	6400
1050	1050	1050	1050	1050	1050	1050	1050	1050	1050
0075	1000	1925	2850	3775	4700	5625	6550	7475	8400
7200	7600	8000	8000	7600	7600	7600	7600	7200	6800
1975	1975	1975	1975	1975	1975	1975	1975	1975	1975
0075	1000	1925	2850	3775	4700	5625	6550	7475	8400
7200	8000	8000	6800	6400	6000	6400	7200	7600	6800
2900	2900	2900	2900	2900	2900	2900	2900	2900	2900
0075	1000	1925	2850	3775	4700	5625	6550	7475	8400
6800	7600	7600	6400	6000	6000	6000	6400	7600	7200
3825	3825	3825	3825	3825	3825	3825	3825	3825	3825
0075	1000	1925	2850	3775	4700	5625	6550	7475	8400
6800	7200	7600	6800	6000	6000	6000	6000	7200	7200
4750	4750	4750	4750	4750	4750	4750	4750	4750	4750
0075	1000	1925	2850	3775	4700	5625	6550	7475	8400
6800	7200	7600	6800	6000	6000	5600	5600	6400	6800
5675	5675	5675	5675	5675	5675	5675	5675	5675	5675
0075	1000	1925	2850	3775	4700	5625	6550	7475	8400
6800	7200	7600	7600	6800	6000	5600	5600	5600	6000
6600	6600	6600	6600	6600	6600	6600	6600	6600	6600
0075	1000	1925	2850	3775	4700	5625	6550	7475	8400
6400	6800	7200	7600	6800	6000	5600	5600	5600	5600
7525	7525	7525	7525	7525	7525	7525	7525	7525	7525
0075	1000	1925	2850	3775	4700	5625	6550	7475	8400
6000	6400	6800	7200	6800	6000	5600	5600	5200	5200
8450	8450	8450	8450	8450	8450	8450	8450	8450	8450
0075	1000	1925	2850	3775	4700	5625	6550	7475	8400
5600	6000	6400	6800	6800	6800	6000	5200	5200	5200

Describe the earth feature you have drawn.

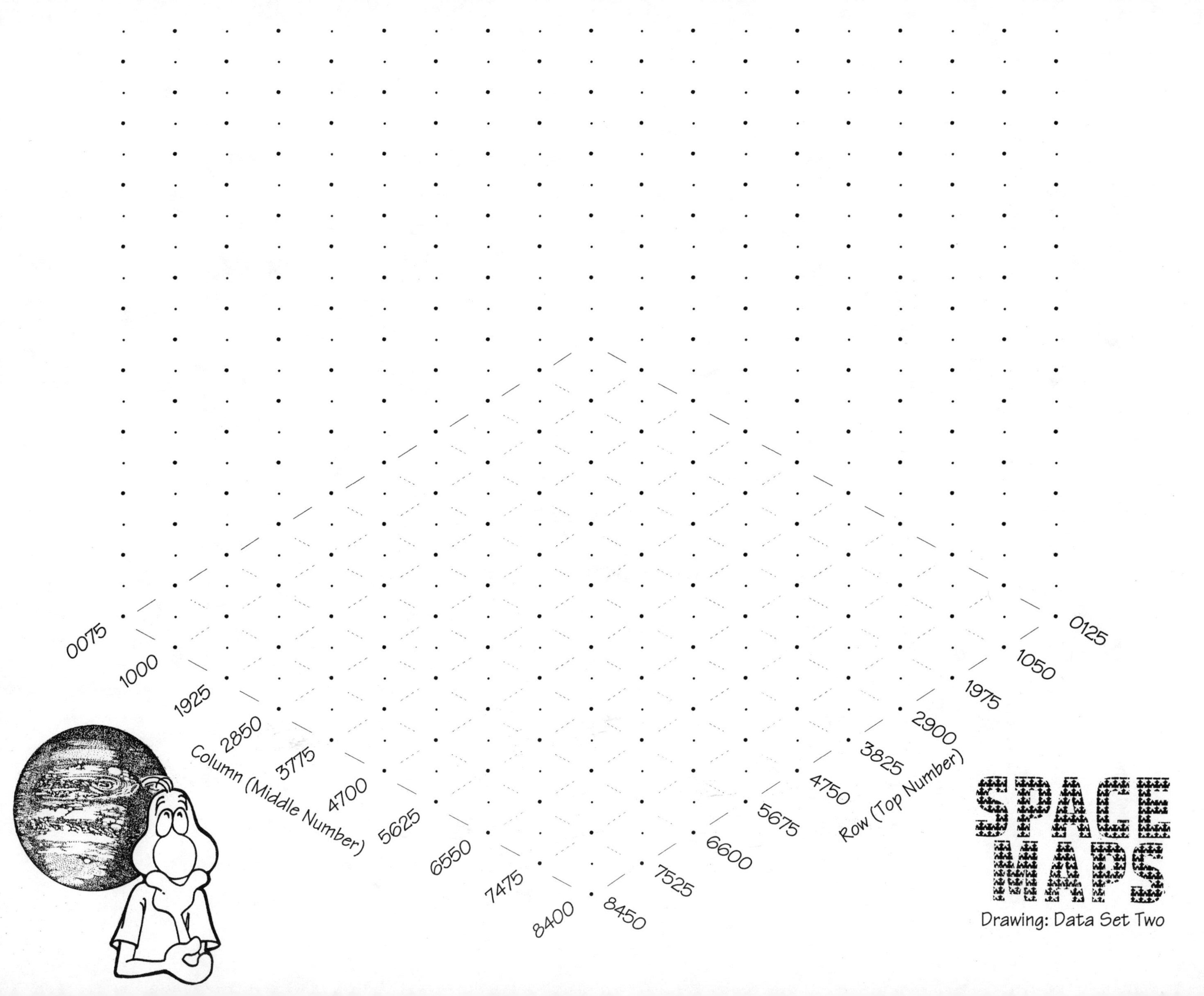
SPACE MAPS
Drawing: Data Set Two
Row (Top Number)
0125
1050
1975
2900
3825
4750
5675
6600
7525
8450
Column (Middle Number)
0075
1000
1925
2850
3775
4700
5625
6550
7475
8400

TOPPING OFF MOUNT SAINT HELENS

Topic
Topographical maps

Key Question
How can the steepness of geological features be shown on a flat map?

Focus
In this activity students will use contour maps to construct models of Mt. St. Helens before and after the 1980 eruption.

Guiding Documents
NCTM Standards

- *Explore problems and describe results using graphical, numerical, physical, algebraic, and verbal mathematical models or representations*

Project 2061 Benchmarks

- *Some changes in the earth's surface are abrupt (such as earthquakes and volcanic eruptions) while other changes happen very slowly (such as uplift and wearing down of mountains).*

Math
Scale drawing
Visualization skills
Topographical mapping

Science
Earth science
geology
volcanoes
mapping

Social Science
Geography
Topographical mapping

Integrated Processes
Observing
Comparing and contrasting
Generalizing

Materials
carbon paper
corrugated cardboard, 26 - 6"x 6" pieces per set of models
scissors or craft knives
glue
optional: paint

Background Information
A contour line on a topographical map shows all the points of equal elevation. If you were to follow one line around a mountain you would remain at a consistent elevation, neither going up or down.

Contour lines which appear to be very close together indicate a rapidly changing elevation. The corresponding area on the mountain will be steep. In contrast, lines which are further apart indicate a change in elevation that is more gradual and the corresponding slope is gentler.

The maps made by the U.S. Geological Survey (from which the maps in this activity have been taken) have been made using a technique called *photogrammetry*. In this technique two photographs are taken by a plane flying over an area. The photographs are very similar, but being taken from slightly different positions they have subtle differences. When the two photos are positioned under a viewer, the area takes on a three-dimensional perspective, much like what you see while looking through a stereoscope. The surveyor uses the three-dimensional perspective to draw in the contour lines.

The eruption of Mount St. Helens on May 18, 1980 lowered the mountain's height by 1,379 feet. The eruption left a crater 2400' deep and displaced 3,179,475,800 cubic yards of volcanic rock (0.5832 cubic miles). The eruption released 1.7×10^{18} joules. This is the equivalent of 27,000 Hiroshima-sized atomic bombs exploding over a nine hour period.

Management

1. Gather necessary cardboard before beginning. If you choose to have each group build a model, you may assign each group to gather their own cardboard. Use a paper cutter to cut it into 6" x 6" pieces.
2. This activity is designed as a class project. Each student is assigned one contour line to trace and cut out. Extra students can be assigned the jobs of gluing and assembly.
3. Craft knives make more accurate and easier cutting of the cardboard. You may find scissors are more easily accessible and less dangerous.

Procedure

1. Discuss the *Key Question* with the class.
2. Distribute student sheets and ask students how the contour lines might describe a mountain.
3. Distribute cardboard, carbon paper, and cutting tools.

4. Assign each student one of the contour lines on one of the maps.
5. Direct students to use the carbon paper to trace their assigned contour line, and the next higher (smaller) contour line onto their cardboard. Be sure students have the carbon side of the paper facing the cardboard, and that the map and cardboard do not shift while they are tracing.
6. Have them cut the cardboard along their assigned contour line.
7. Glue the cardboard pieces together starting with the lowest contour piece. Glue on the next lowest piece using the carboned line on the lowest piece for placement. Continue gluing the pieces in sequence until you get to the highest piece.
8. When the models are dry, students may choose to paint them.

Discussion

1. How does the model compare to a picture of Mt. St. Helens?
2. How do the two models differ? [crater, height]
3. How do these differences show up on the map? [indentation of contour lines, less contour lines on map of shorter mountain]
4. Look at the steep sides of the crater on the second model. How does this steepness show up on the map? [contour lines are very close together]
5. Using the scale on your map and looking at the model mountains, estimate how much rock was moved to make the crater. [Answers will vary, but discuss strategies for estimating. You might take modeling clay to make the top of the original mountain on the crater of the new mountain, and then reform the clay into a cube and scale an edge. See *Background Information*.]

Extensions

1. Have students look at actual topographical maps and describe the lay of the land.
2. Get a topographical map of a local area and make a model using it.

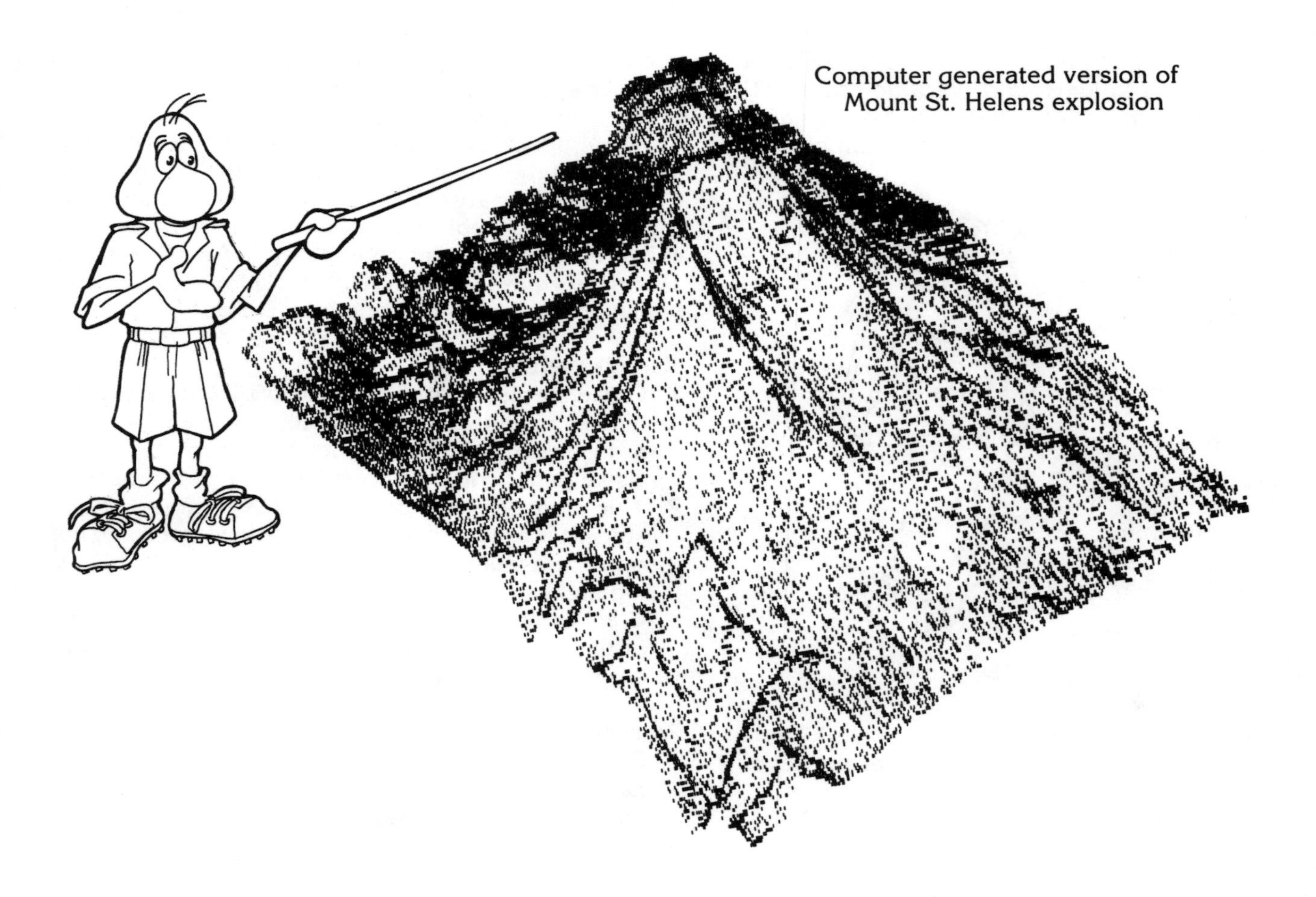
Computer generated version of Mount St. Helens explosion

TOPPING OFF MOUNT SAINT HELENS

Before the 1980 Eruption

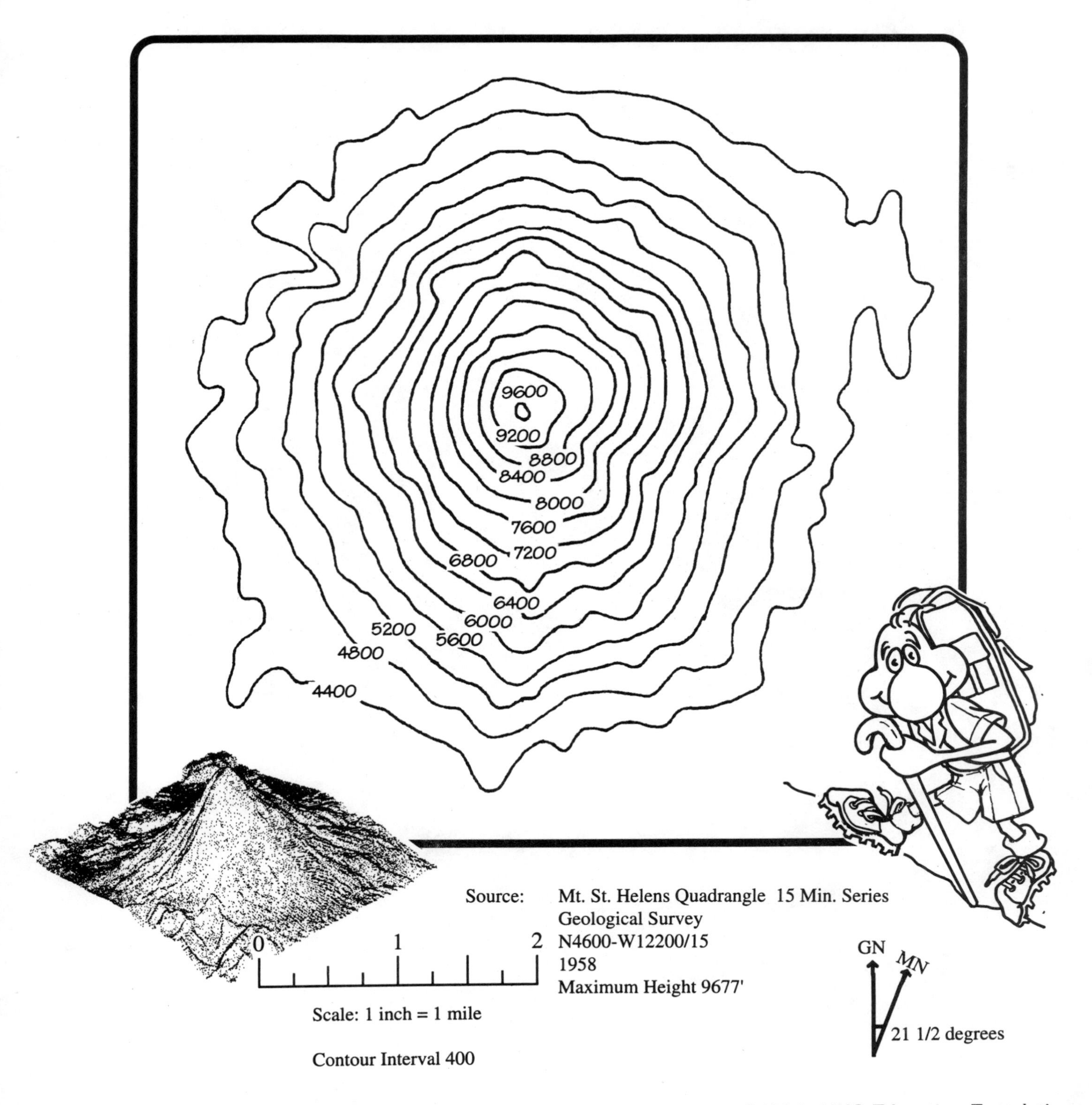

Source: Mt. St. Helens Quadrangle 15 Min. Series
Geological Survey
N4600-W12200/15
1958
Maximum Height 9677'

Scale: 1 inch = 1 mile

Contour Interval 400

TOPPING OFF MOUNT SAINT HELENS

After the 1980 Eruption

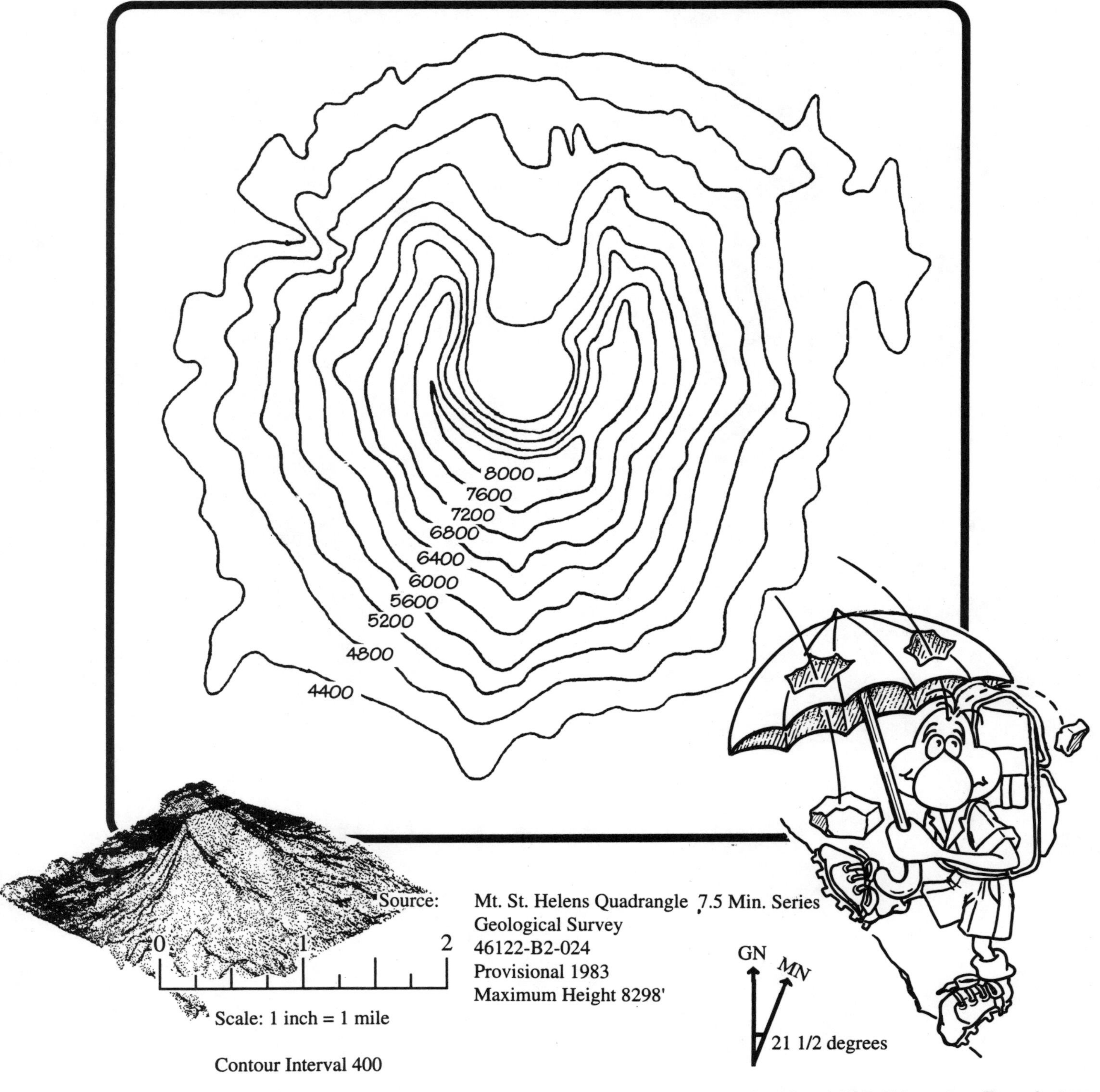

UNCANNY VISION

Topic
Making contour maps

Key Question
How can you determine what is inside this can without opening it or shaking it?

Focus
Students will make a two-dimensional contour map of an unknown three dimensional object.

Guiding Documents
NCTM Standards
- *Develop and apply a variety of strategies to solve problems, with emphasis on multistep and nonroutine problems*
- *Understand and apply reasoning processes, with special attention to spatial reasoning and reasoning with proportions and graphs*
- *Make inferences and convincing arguments that are based on data analysis*

Project 2061 Benchmarks
- *Mathematics is helpful in almost every kind of human endeavor–from laying bricks to prescribing medicine or drawing a face. In particular, mathematics has contributed to progress in science and technology for thousands of years and still continues to do so.*
- *Technology is essential to science for such purposes as access to outer space and other remote locations, sample collection and treatment, measurement, data collection and storage, computation, and communication of information.*

Math
Measuring
Estimating
Graphing
Interpolating
Spatial visualization

Science
Earth science
  geology
  oceanography
  topographical mapping

Social Science
Geography
  longitude, latitude
Topographical mapping

Integrated Processes
Observing
Predicting
Collecting and recording data
Interpreting data
Inferring
Applying

Materials
empty coffee cans with plastic lids
bamboo skewers
student sheets
rulers
objects for mapping (cubes)
clay
glue or tape
marker
nail
hammer

Background Information
Contour maps of submerged objects are made by soundings. Modern techniques use sonar (SOund NAvigation Ranging) to make measurements. Sound is reflected off the submerged objects and the time for the sound to travel is converted into depth. This activity uses a more primitive method of physical soundings. A probe is submerged until it comes in contact with the bottom. The depth is then directly measured from the length of the probe. When the soundings are recorded in the corresponding positions where they were made, then the points of equal depth can be connected with a line. These lines form contour lines. (For more information on contour lines refer to *Background Information* in *Topping Off Mount St. Helens*.)

Management
1. Before beginning, glue or tape the appropriate grid from the student sheets to the lid of the coffee can. With the nail and hammer, make a hole through the grid and the lid at each intersection on the grid. Secure an object to the bottom of the can with modeling clay. Place the lid on the can and align the grid's *north* with the seam of the can.

2. The first time students do this activity you should use a simple object. Mounds can be easily made from plastic or wood cubes. As the students gain skill, more difficult objects may be substituted.
3. This activity may be done with several students at a center or as a class activity with students taking turns sounding and all students recording. It may be done as a small-group activity if enough cans are gathered beforehand. Combining these methods by initially working as a class and then splitting up into small groups for mapping a second object also works quite well.
4. A mapping grid is provided for the three common sizes of coffee cans. Have students make their records on the same-size grid as the can being used. This will allow students to make a direct comparison of their map to the object.
5. Students may use several ways to measure and map the object inside the can. The first one is to measure the distance the skewer goes down until it touches the object. This method is the one followed in the *Procedure*. Another way is to measure the height of the skewer above the lid when the skewer is touching the object. This method produces a map that shows the relative shape of the object, but it does not show the object's depth in the can. The third way is to stick a skewer in every hole. This produces a physical display of the shape of the object. This final approach provides a very effective demonstration of how soundings work. To produce a map, each skewer still needs to be measured.

Procedure

1. Discuss the *Key Question* with the class while focusing their attention on the can.
2. Distribute materials: can with object and grid, skewer, ruler, student record sheet.
3. Instruct students to make a sounding for each hole by lowering the skewer in each hole until it touches the bottom. Emphasize to the students that they must keep the skewers perpendicular to the bottom of the can.
4. At each hole, have the students mark the height of the skewer where it comes through the lid, remove the skewer, measure the depth at that point (to the nearest centimeter), and record the depth on the corresponding intersection on the grid on their record sheet.
5. When all the soundings are complete, have students draw in the contour lines by connecting points of equal depth. It is easiest if students first connect the points of greatest depth and then work up to the points of least depth. The students may need to discuss the situation where there is a gap between the depths of contour lines. (If a contour line of 12 cm and contour line of 14 cm are next to each other, then somewhere between those lines a contour line of 13 cm needs to be drawn, even though it was not measured by the soundings.)
6. Using their completed maps, direct the students to write a description of what they expect their object(s) to look like.
7. Have students take the lid off the can and compare their predictions from the map with the actual object.

Discussion

1. How do you know a high or low spot on the map? [depth the skewer goes into the can]
2. How can you tell a steep or level area? [close contours lines indicate steep areas, widely-spaced contour lines indicate more level areas]
3. How accurately does your map represent the object? (answers will vary)
4. How could more accurate maps be made? [more accurate measurement on soundings (mm); use a smaller grid]

UNCANNY VISION

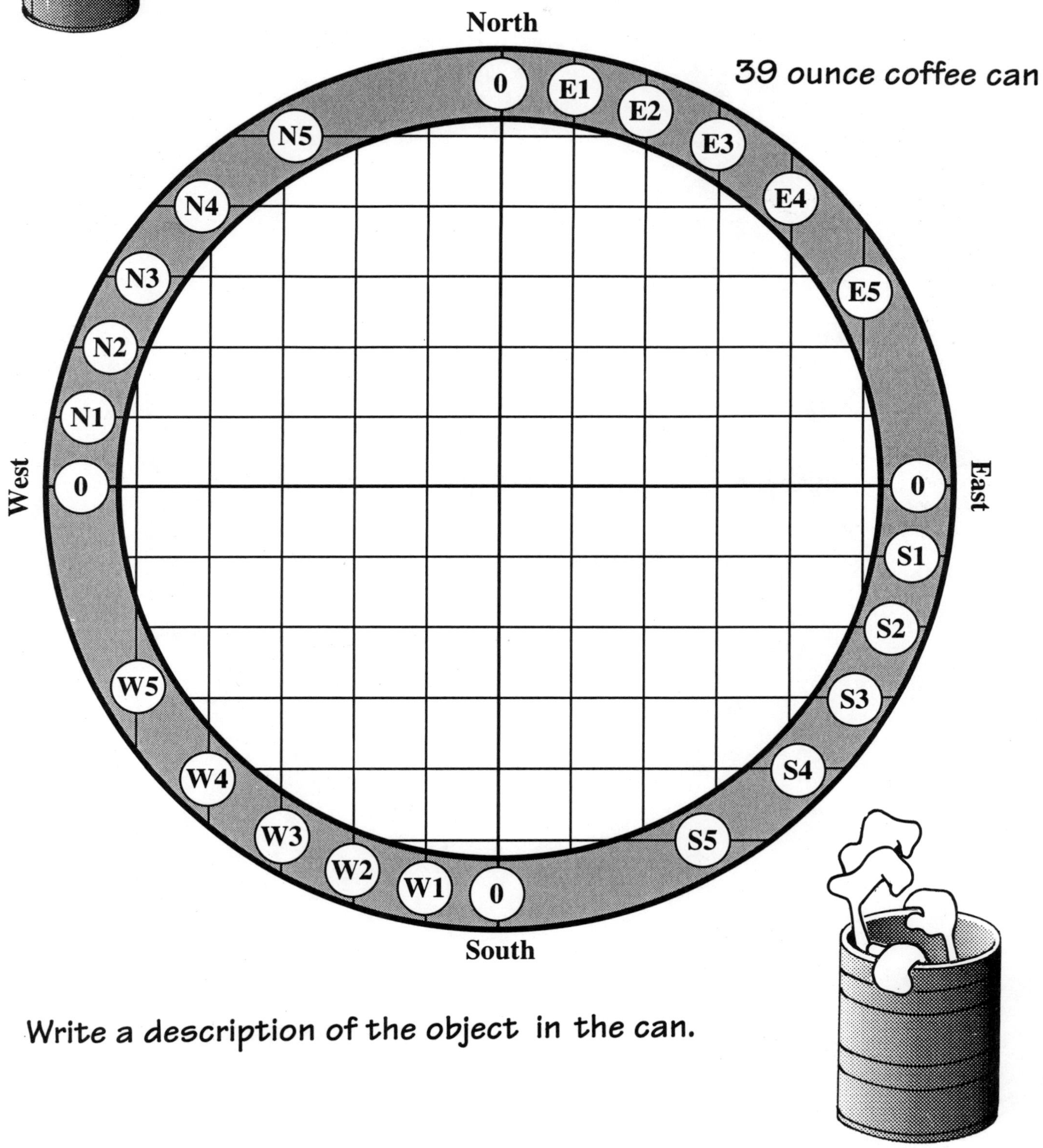

Write a description of the object in the can.

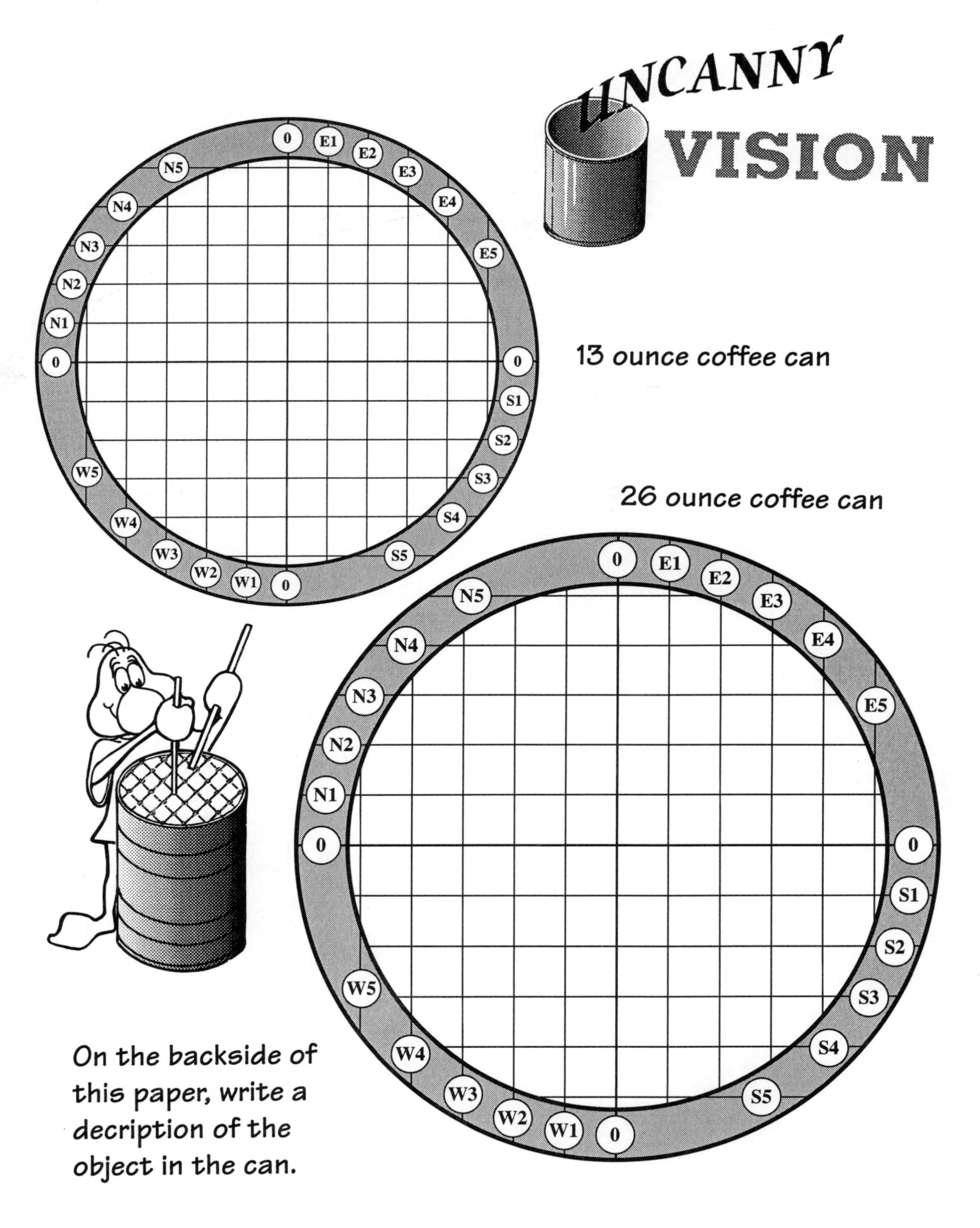
UNCANNY
VISION
0
E1
E2
E3
E4
E5
N5
N4
N3
N2
N1
0
0
S1
S2
S3
S4
S5
W5
W4
W3
W2
W1
0
13 ounce coffee can
26 ounce coffee can
0
E1
E2
E3
E4
E5
N5
N4
N3
N2
N1
0
0
S1
S2
S3
S4
S5
W5
W4
W3
W2
W1
0
On the backside of this paper, write a decription of the object in the can.

On The Level

Topic
Topographical mapping

Key Question
If you needed to make your yard level, how could you determine where the high and low spots were?

Focus
Students make a topographical map of a small area using a level and tape measure.

Guiding Documents
NCTM Standards
- *Develop and apply a variety of strategies to solve problems, with emphasis on multistep and nonroutine problems*
- *Generalize solutions and strategies to new problem situations*
- *Develop number sense for whole numbers, fractions, decimals, integers, and rational numbers*

Project 2061 Benchmarks
- *A number line can be extended on the other side of zero to represent negative numbers. Negative numbers allow subtraction of a bigger number from a smaller number to make sense, and are often used when something can be measured on either side of some reference point (time, ground level, temperature, budget).*
- *Organize information in simple tables and graphs and identify relationships they reveal.*
- *The graphic display of numbers may help to show patterns such as trends, varying rates of change, gaps, or clusters. Such patterns sometimes can be used to make predictions about the phenomena being graphed.*

Math
Measurement
Integers
Scale drawings
Spatial visualization

Science
Earth science
  geology
  topographical mapping

Social Science
Geography
  topographical mapping

Integrated Processes
Observing
Comparing and contrasting
Collecting and recording data
Interpreting data
Applying

Materials
Meter sticks
1/4" vinyl tubing (aquarium tubing available at pet stores, 12' minimum per group)
Tape measure
Masking tape
Paper cups
Food coloring
String
Wood sticks (see *Management 2*)
Student sheets

Background Information

The level in this activity is made from vinyl tubing filled with water. The last meter of both ends of the tube is taped to meter sticks. Both meter sticks and their attached tubes are placed next to each other in a vertical position on a table. Water is then put in the tube until the water level reaches about 50 cm. The height of water will be the same in both tubes.

This level can now be used to tell the relative altitude of any two positions by the change in the water's height. For example, one meter stick remains on a table and the other stick is lowered 30 cm to the seat of a chair. The water level in the meter stick on the table will be at the 35 cm and the water level in the meter stick on the chair will be at 65 cm. The difference in the scale readings is 30 cm, the difference in the heights of the meter sticks' positions. The chair is lower than the table but the height of the water in that stick has a higher scale reading. The positions' levels are inverse of the scale reading: the meter stick with the higher scale reading is the lower stick.

Altitudes are recorded relative to an origin. Customarily, sea level is the origin; however, the origin may be any arbitrary point. To determine the altitude of a remote point, the changes in altitudes to all the points in-between the origin and the desired position are sequentially measured. If the next position from a point is higher, a positive value is recorded. If the next position is lower, a negative value is recorded. The altitude of any position in relation to the origin is the sum of all the altitude changes before it in the sequence. By using a running total, the altitude relative to the origin is always known.

A topographical map of a small area can be made by dividing the area up using a grid. An intersection on the grid is established as the origin. Using a tape measure to identify the other intersections, the level is used to establish each intersection's relative altitude to the origin. The relative altitudes are recorded on the corresponding intersections of a grid on a scale drawing. The intersections of the same altitude are connected with a contour line and the map is formed.

Management

1. Before beginning this activity, the teacher needs to find an area that is appropriate to map. The area should approximate a typical back yard (15-20m x 10-15m). It should not be flat. The greater the differences in altitudes, the better–as long as the differences are not greater than one meter of rise for each meter of run (horizontal distance).
2. If the teacher will mark out this area before doing *Part 2,* it will alleviate a lot of confusion. Start by surrounding the rectangular area to be mapped with a string. Tie the string to sticks stuck in the ground to serve as corners. Make sure the rectangle has right angles and opposite sides of equal lengths. Tape strips of paper which have been labeled alphabetically at one meter intervals to the string along one side of the rectangle. Put numbered markers along the opposite side of the rectangle. Designate the *A,1* marker as the origin.
3. This activity will take about three to four periods. One or two periods will be used for *Part 1*, determining how to measure with the level. Another period will be used to take the measurements and determine relative altitudes (*Part 2*). The last period will be used for *Part 3-Making the Map.*
4. This activity is written for small groups or three to four students. If materials are in short supply, one level can be constructed and students can take turns making measurements and sharing information with the class.

Procedure

Part 1 - Making and Using the Level

1. Discuss the *Key Question* with the class.
2. Distribute two meter sticks, tubing, tape, and *Reading the Level* sheet to each group.
3. Have students construct the level as explained on instruction sheet.
4. Ask students to come up with a method of how they can measure the difference in the height of two points. If this open-ended approach is not appropriate for the class, the second instruction sheet may be used to provide more structure.
5. Have students share their observations and make suggestions of how they might use the level to solve the *Key Question.*

Part 2 - Making Measurements of Area

1. Distribute materials for mapping: level, tape measure, and record sheet.
2. Discuss with students the general method of mapping (see *Background Information*) and how to use the record sheet to determine relative altitude.
3. Take the class out to the area to be mapped.
4. If not previously done, define a rectangular area with string.Tie the string to sticks stuck in the ground to serve as corners. Make sure the rectangle has right angles and opposite sides of equal length.
5. Tape a marked strip of paper at one meter intervals to the string along one side of the rectangle. Label the strips alphabetically starting with *A* in one corner. Put numbered markers along the opposite side of the rectangle.
6. Designate the *A,1* marker as the origin.
7. As a class, use the level to establish the relative altitudes of all the lettered positions on the origin's side. This side is referred to as the baseline. The students record the measurements on the *Baseline Measurement* chart and calculate the relative altitude.
8. Assign each group a numbered line to measure going across the rectangle from each marker.
9. Have each group stretch a tape measure between its two corresponding markers, one on the baseline and one on the opposite side.
10. Direct students to place one end of the level at their assigned marker on the baseline of which the relative altitude is known, the other end is placed one meter along the tape measure from this baseline position.
11. Have students use the water level to determine the relative altitude of the new position, recording and calculating it on the *Line ____ Measurement* chart.
12. Direct them to move the level so one end is at the position where the altitude was just determined and the other end is one more meter down the tape measure.
13. Have them measure and record the altitude of this new position.
14. Instruct them to continue to move across their assigned lines as directed in steps 12 and 13 until they reach the other side.

Part 3 - Making the Map

1. Distribute the *Level Lots Mapping, Inc.* sheet.
2. Have each group report the relative altitudes they measured. All students should record the relative altitudes on the corresponding positions on the grid.
4. Direct them to draw lines to separate altitudes of given ranges (0-10 cm, 10-20 cm, 20-30 cm, etc.). These lines of separation are contour lines.
5. Discuss the maps.

Discussion

1. Describe what a trip down your assigned line would be from an ant's perspective.
2. What features were in the area you mapped?
3. How do the features of the area show up on the map?
4. Where is the lowest point on the map?
5. Where is the highest point on the map?
6. Where is the most level area on the map?
7. Where is the steepest slope on the map?
8. How well does this map represent the area you mapped?
9. How could you improve the accuracy of the map?

Extensions

1. Have each group of students draw and cut out from card stock a scaled cross-section of their assigned line. Attach the cross-sections to cubes so that they stand up vertically. When all the groups' cross-sections are sequenced and put next to each other they will form a model of the area.
2. Have students determine a way to use a garden hose as a level to make the top of two remote sticks level.

On The Level

Level Instruction Sheet

Making the Level

1. Place one end of the tube in half a cup of water and sip water through the tube until all but the last meter of tubing is filled with water.

2. Keeping the ends of the tube level with each other so no water is spilled out, tape in several places the last meter of plastic tubing to one of the meter sticks. Make sure the end of the meter stick with the 100 cm mark is next to the open end of the tube.

3. Do the same thing with the other end of tubing and the other meter stick. Again make sure the 100 cm mark is next to the open end of the tube.

4. Place both meter sticks upright next to each other. Adjust the amount of water until the water level in both tubes is close to 50 cm.

How could you use this tool to determine if two positions were level?

How could you use this to determine how much higher or lower one is from another position?

READING THE LEVEL

Learn how to read the level by doing the following:

1. Place the bottoms of both meter sticks on the same table.

2. Lower one of the meter sticks 45 cm. The top of the moved meter stick will be at the 55 cm mark on the stationary meter stick.

3. Record on the chart the water level of the water on both meter sticks.

4. Move the meter stick up to complete all the observations on the chart.

Water Level in Moved Meter Stick							
Water Level in Stationary Meter Stick							
Amount to Raise (+) or Lower (-) Meter Stick	-45	-30	-15	0	+15	+30	+45

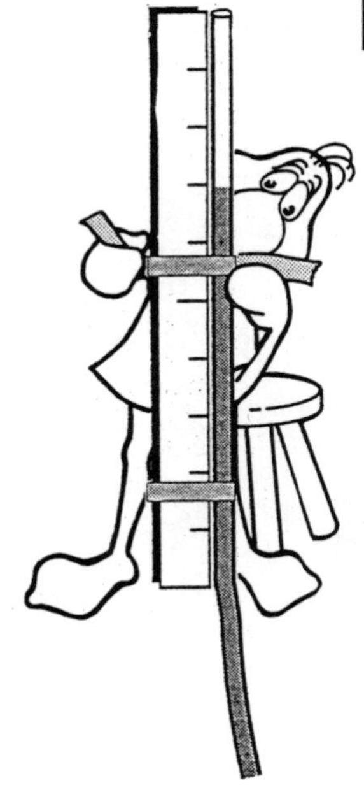

From observing the water levels, how could you determine how much higher one meter stick is than the other?

From observing the water levels, how can you tell if the meter stick you moved is higher or lower than the stationary one?

How could you use this level to determine the difference in heights of two places?

On The Level

Baseline Measurements

Coordinate Position of Point Being Measured	Water Level on Scale at Known Point	Water Level on Scale at Point Being Measured	Difference in Water Levels at Both Points	Relative Altitude from Known Point (+=higher, -=lower)	Relative Altitude of Point Being Measured to Origin

Line ___ Measurements

Coordinate Position of Point Being Measured	Water Level on Scale at Known Point	Water Level on Scale at Point Being Measured	Difference in Water Levels at Both Points	Relative Altitude from Known Point (+=higher, -=lower)	Relative Altitude of Point Being Measured to Origin

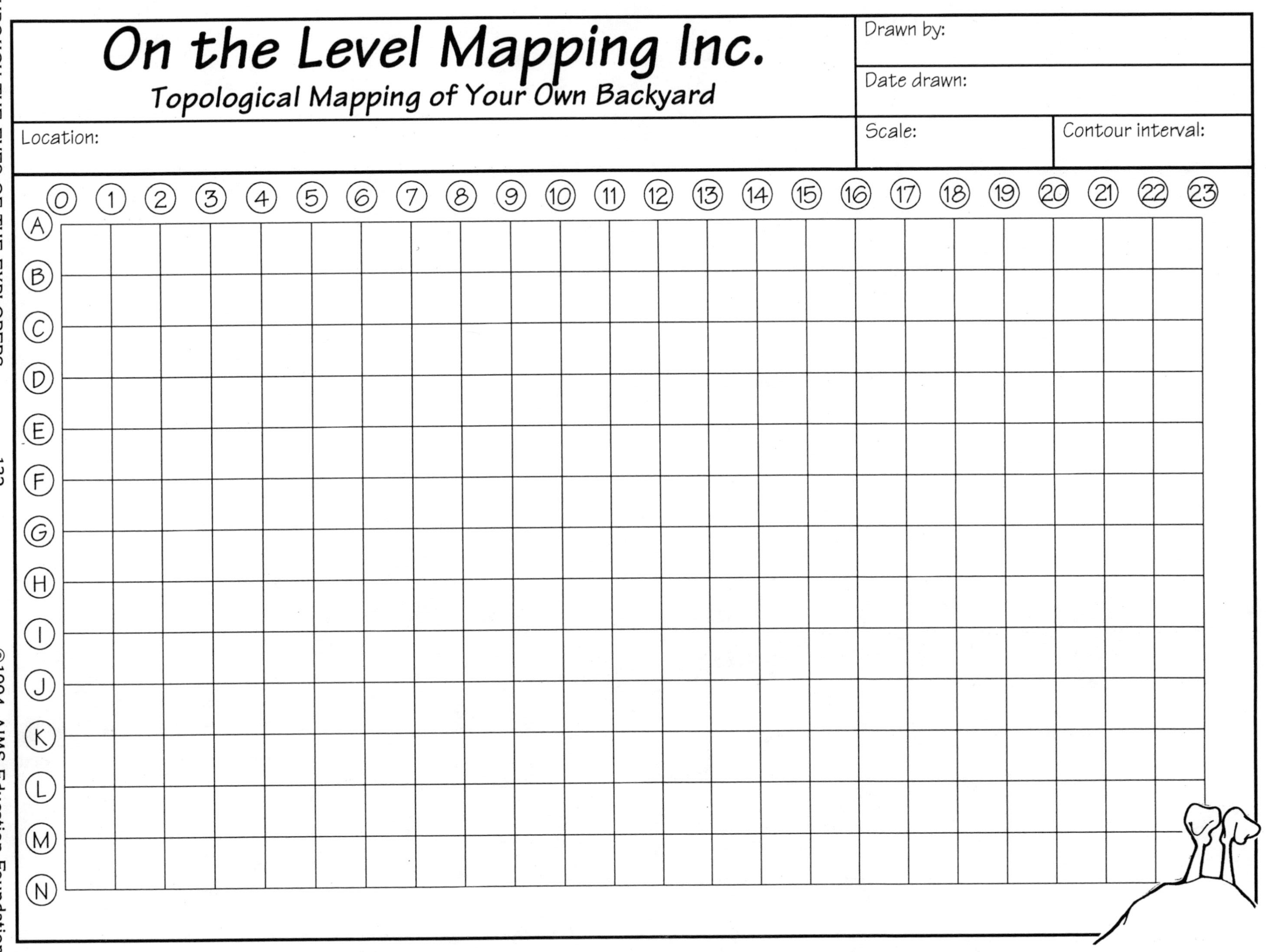

On the Level Mapping Inc.
Topological Mapping of Your Own Backyard
Drawn by:
Date drawn:
Location:
Scale:
Contour interval:
0 1 2 3 4 5 6 7 8 9 10 11 12 13 14 15 16 17 18 19 20 21 22 23
A B C D E F G H I J K L M N

HILL-SIDES

Topic
Topographical map of a cross section of a hill

Key Question
If you were going to make a road across a hill, how could you make a picture of how steep it was going to be?

Focus
Students will take field measurements and make a scale drawing of a cross section of a hill.

Guiding Documents
NCTM Standards
- *Compute with whole numbers, fractions, decimals, integers, and rational numbers*
- *Understand, represent, and use numbers in a variety of equivalent forms (integer, fraction, decimal, percent, exponential, and scientific notation) in real-world and mathematical problem situations*
- *Estimate, make, and use measurements to describe and compare phenomena*

Project 2061
- *The scale chosen for a graph or drawing makes a big difference in how useful it is.*
- *Organize information in simple tables and graphs and identify relationships they reveal.*
- *Numbers can be written in different forms, depending on how they are being used. How fractions or decimals based on measured quantities should be written depends on how precise the measurements are and how precise an answer is needed.*
- *The expression a/b can mean different things: a parts of size 1/b each, a divided by b, or a compared to b*

Math
Measurement
Geometry
Ratio and proportion
Scale drawings
Integers
Graphing
Slope
Spatial visualization

Science
Earth Science
geology
mapping

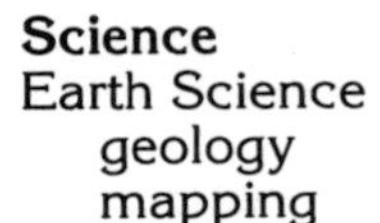

Social Sciences
Geography
mapping

Integrated Processes
Observing
Collecting and recording data
Interpreting data
Applying

Materials
Meter sticks
Straws
String
Tape
Paper clips
Tape measure
Strips of bright paper
Student sheets

Background Information
A vertical cross-sectional drawing of a hill shows how it would look if it were cut vertically along a line. The drawing would show how the slice would look if viewed from one side.

To get the measurements to make a cross section, a device that allows one to sight a level line is used. The level-sighting device is placed one meter above the ground at the bottom of a line along which the cross section is being made. When the sighting device is level, a line is sighted through a straw up the cross section line. The point where the line of sight hits the hill is marked. The horizontal interval from the top of the sighting device to the marked point (one meter of elevation) is measured. The sighting device is then moved to the marked point and the next vertical interval of a meter is sighted.

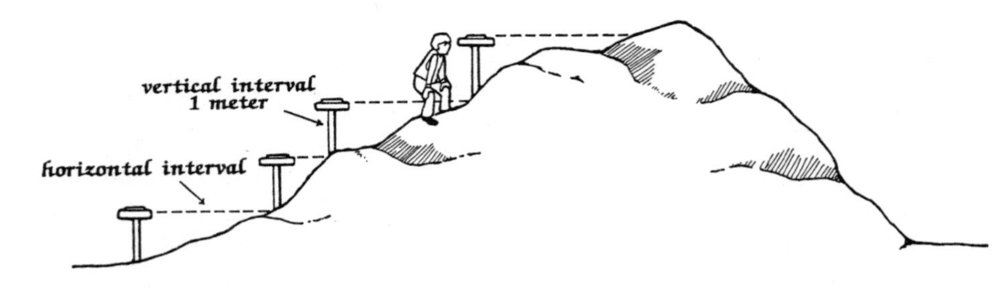

To make a sighting on a slope that goes down, the process is reversed. The sighting device is moved along the cross-section line until, when you look back, the last sighted point is level with the device. The point at the sighting device becomes the marked position.

After a series of these sightings has been made, a graph of the horizontal and vertical intervals can be constructed. When these points that represent the marked sighting points are connected, a cross section of the hill evolves. It will be a true representation of the hill if both horizontal and vertical scales are the same. In a cross section, the vertical scale is often exaggerated. This is easily done by using a larger scale on the graph for the vertical scale than the horizontal scale.

The information gathered can be used to determine the slope of any part of the hill or average of the whole hill. Slope is the comparison of the vertical interval to the horizontal interval. Slope may be expressed as a ratio, a decimal, or a percent. Slopes with positive values are going up; slopes with negative values are going down. A slope with a value of 1, or 100%, is going up at a 45° angle. Any hill with a slope of less than 1, or 100%, is less steep that a 45° angle. Mathematicians use the fraction or decimal notation for slope. The construction industry uses the percent notation or feet of rise per 100 feet of horizontal run to describe slope. (The interstate highway system has been designed with a maximum slope of 6%. The steepest grade on a main railroad line is 2.8%.)

Management

1. An appropriate slope is necessary for this activity. There must be a rise of at least 3-4 meters. With the tools being used, it is difficult to be accurate on any slope of less than 5% (1 m vertical/20 m horizontal).
2. The teacher may choose to build the *Level Sights* before beginning the activity to make sure the tool is assembled correctly.
3. Students will need to work in groups of at least three: One to sight through the straw, one to use the plumb line to make sure the sighting device is level, and the third to mark the level point on the slope.
4. Students may need some assistance in making sure their cross-section line is straight. Their sighted points tend to form a less-than-straight line. Encourage students to use a distant landmark as a marker for their cross-section line. You might provide them with fluorescent flagging or bright strips of paper to mark their sighted points. This will allow them to more easily see if they are maintaining a straight line.
5. This activity is the first in a series of two on making a topographical map of a mountain. *Major Mountain Mapping* will explain how to arrange the cross-section lines so a map with contour lines can be made easily. This activity may be taught as a separate activity or as a lead into *Mapping Mountains.*

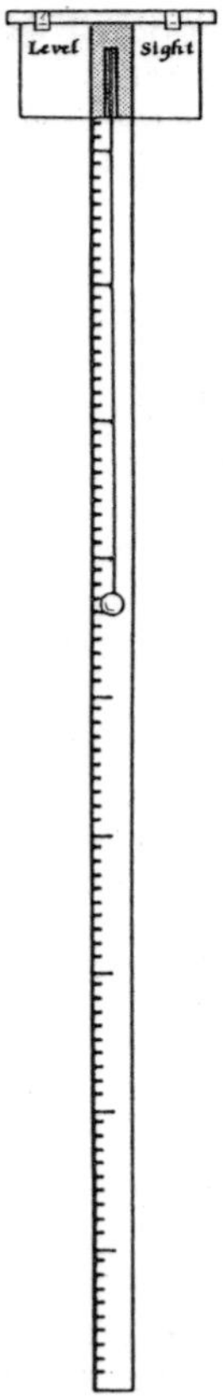

Level Sight

Procedure

1. Discuss the *Key Question* with students.
2. Distribute materials and assemble *Level Sights.*
3. Distribute materials for mapping and go to sighting area.
4. Assign each group a starting position and a cross-section line.
5. Inform each group of the following procedure they will use to measure their line.
   a. Place *Level Sight* at position.
   b. Make a level sight up the cross-section line.
   c. Mark the position where the sighting line hits the slope.
   d. Measure the distance from the top of the *Level Sight* to the marker's position.
   e. Record the horizontal and vertical interval measurements on the record sheet.
   f. Move the *Level Sight* to the most recently marked position.
   g. Repeat steps a-f until measurements are made along the full length of the cross-section line.
6. Have students complete the record sheet by calculating the slopes in fraction, decimal, and percent formats.
7. Guide students in making a cross-sectional profile of the measured area using the following procedure.
   a. Place a dot in the lower left side of the drawing to represent the original assigned position.
   b. Starting at the most recently drawn dot, move the scaled horizontal and vertical distances for the next sighting.
   c. Place a dot at this point to represent this position.
   d. Repeat steps b and c until all the sightings have be mapped.
   e. Connect all the plotted points in sequence to form the profile.

Discussion

1. Where is the steepest part of the profile? What is that part's calculated slope?
2. What is the most level part of the profile? What is that part's calculated slope?
3. Do you have any part of the profile that is lower than the section before it? What is its calculated slope?
4. How does the calculated slope relate to the hill in the profile? [greater slope = steeper hill, positive slope is uphill, negative slope is downhill]
5. What would a hill with a 100% grade look like? [a 45 degree slope]

Extensions

1. To make a three-dimensional model of the hill, have students cut out their cross sections from tagboard. Stand the cross sections vertically and orient them as they were made in the field.
2. Have students use a topographical map and make a cross section of a trip through the mountains on a road or trail.

Hill-Sides

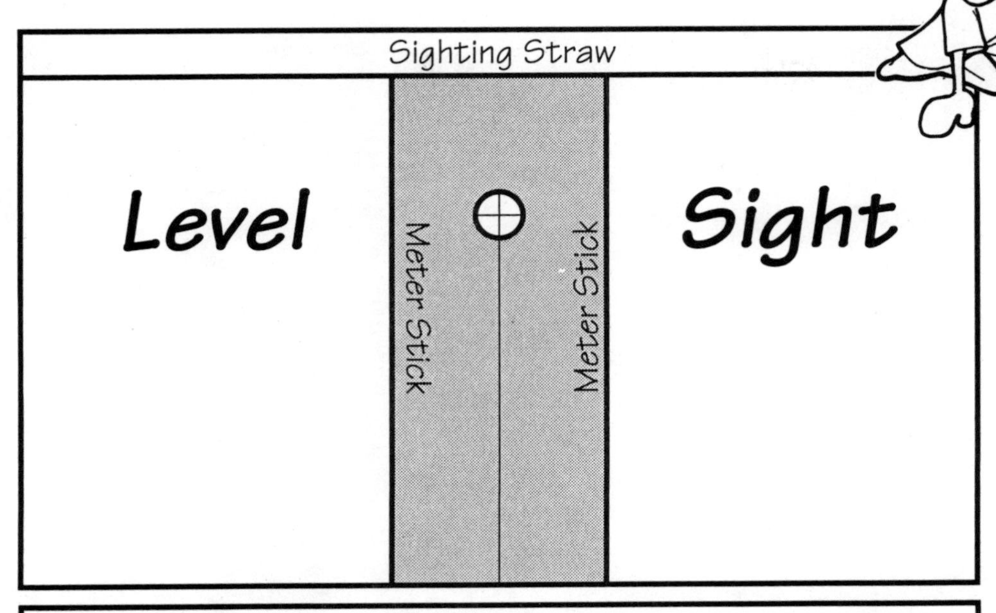

Sighting Straw

Level Meter Stick Meter Stick Sight

Sighting Straw

Level Meter Stick Meter Stick Sight

Assembly Instructions

1. Copy the *Level Sights* onto tagboard and cut them out.
2. With a push-pin make a hole at the cross hairs in the circle.
3. Thread a 50 cm string through the cross hairs and tape the short end onto the back side of the tagboard. Leave the remaining length on the front side.
4. Tape a penny to the long end of string. The string and penny form a plumb line.
5. Tape a straw in the labeled position.
6. Place the *Level Sight*, at the top of a meter stick. The straw must be even with one end of the stick and perpendicular to the stick. The long sides of the shaded region need to be parallel with the sides of the meter stick. Tape the sight in this position.
7. To use the *Level Sight*, place the meter stick vertically with the end opposite the sight on the ground. Move the meter stick until the plumb line is centered on top of the center line on the sight. The line sighted through the straw is level to the ground.

HILL-SIDES

Field Measures			Calculated Slope		
Sighting Number	Horizontal Interval	Vertical Interval	Ratio: Vertical/Horizontal	Decimal Ratio	Percent Slope
1					
2					
3					
4					
5					
6					
7					
8					
9					
10					
11					
12					
13					
14					
15					
16					
17					
18					

Field Measures			Calculated Slope		
Sighting Number	Vertical Interval	Horizontal Interval	Ratio: Vertical/Horizontal	Decimal Ratio	Percent Slope

Hill-Sides

Drawn by:

Date drawn:

Location:

Horizontal scale:

Vertical scale:

MAJOR MOUNTAIN MAPPING

Topic
Topographical map

Key Question
How would you go about making a topographical map of a real mountain?

Focus
A culminating event in which students use a combination of the skills they have learned in other activities to make a topographical map of a mountainous area.

Guiding Documents
NCTM Standards
- *Develop and apply a variety of strategies to solve problems, with emphasis on multistep and non-routine problems*
- *Estimate, make, and use measurements to describe and compare phenomena*
- *Generalize solutions and strategies to new problem situations*
- *Investigate relationships among fractions, decimals, and percents*

Project 2061 Benchmarks
- *Numbers can be written in different forms, depending on how they are being used. How fractions or decimals based on measured quantities should be written depends on how precise the measurements are and how precise an answer is needed.*
- *Read simple tables and graphs produced by others and describe in words what they show.*
- *The expression a/b can mean different things: a parts of size 1/b each, a divided by b, or a compared to b.*

Math
Measurement
Geometry
Ratio and proportion
Scale drawings
Spatial visualization

Science
Earth science
  geology
  topographical mapping

Social Science
Geography

Integrated Processes
Observing
Collecting and recording data
Interpreting data
Analyzing
Applying

Materials
For each group:
  magnetic compass
  tape measure
  level sight (see *Hill-Sides*)
  florescent flagging or strips of bright paper

For the class:
  ruler
  protractor
  water level (see *On the Level*)
  vellum

Background Information

The distance between meter intervals of altitude was gathered in making cross sections in the activity *Hill-Sides.* These distances provide the necessary measurements for making a contour map of an area. The cross sections must be oriented in relationship to each other.

The appropriate orientation of the cross-section lines is determined by the lay of the land in the area being mapped. If the area has a definite peak, the lines should radiate from this peak. If the area is a valley encircled by hills, the lines should radiate from the low center of the area. Both of these orientations have a point common to all the lines. The relative altitudes along all the lines are made from this common point. The maps made from a central point of orientation are the easiest to make and are therefore an appropriate starting point.

If the area to be mapped is a long ridge, the cross-section lines will be parallel to each other and will lie perpendicular to a baseline at regular intervals. First a straight baseline is sighted through the area. This line may lie atop the ridge or along the base. At regular intervals along this line, markers are left to indicate cross-section lines. Using a level sight (see *Hill-Sides)* or water level (see *On the Level*), determine the relative altitude of each marker on the baseline to the nearest meter. Each group then measures a cross-section line from each marker that runs perpendicular to the baseline as done in *Hill-Sides*. The orientation is maintained using a compass to keep the cross-section line on the correct heading. The relative altitude is kept from the common point or the origin on the baseline.

The cross sections are oriented and scaled using a ruler and protractor. The altitudes are then placed on the scale map of the area. Points with the same relative altitude are connected to form contour lines.

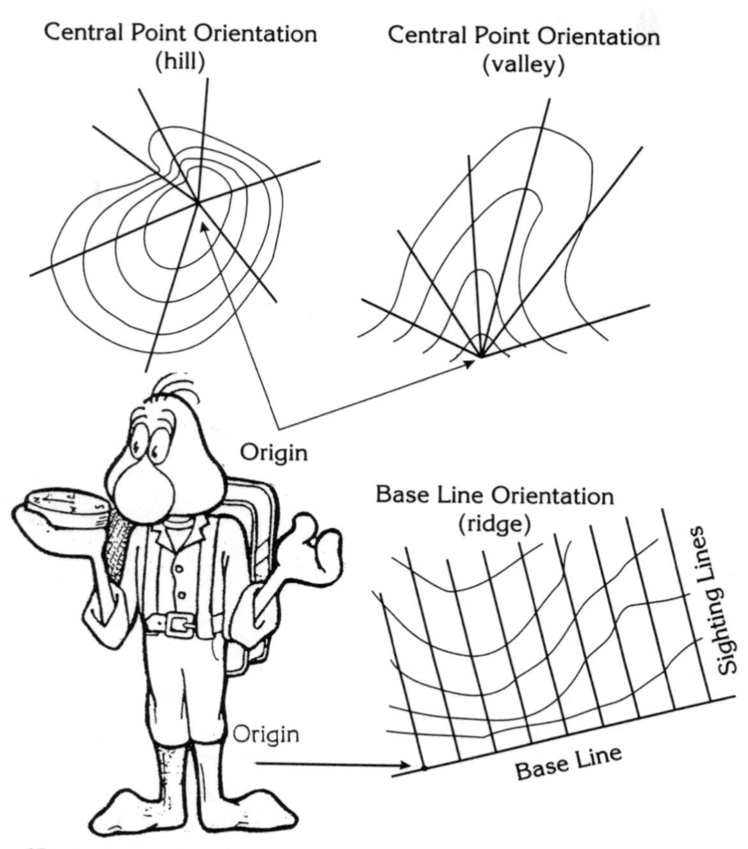

Management

1. Before doing this activity find a suitable area to map. A good-sized hill with some steep terrain will make an exciting challenge. Avoid hills covered with thick undergrowth or buildings. Open woods and grassy hills are excellent.
2. The amount of time allotted for the activity is dependent on the size of the area to be mapped. Several hours are required to complete the field measurements. Prior to doing the activity, the teacher can make a trip to the area to do the *Setting up the area* section of the procedure to shorten the time.
3. When students are sighting their lines, encourage them to use a distant landmark as marker. Provide them with florescent flagging or bright strips of paper to mark their lines. Greater accuracy may occur if students do their sightings to mark the line, then go back to find their interval measurements.
4. Before attempting this activity, make sure your students have mastered the skills in *Treasure Hunt* and *Hill-Sides*.
5. Vellum is large graph paper used for drafting. It is available in stationery stores and art supply stores.

Procedure

Setting up the area

1. Go to the area and determine what type of cross-section line orientation is most appropriate for the lay of the land.

If the central-point orientation is appropriate

2. Determine and identify the central point.
3. Determine the compass headings for each cross-section line. Attempt to send a cross-section line through each major land form (gully, ridge, outcropping).
4. Assign compass headings to each group for mapping.

If the baseline orientation is appropriate

2. Sight and mark baseline along one heading of the compass. It helps to choose a baseline that is relatively level. In a park, sidewalks often provide good baselines.
3. Mark the baseline at regular intervals for cross-section lines. (10 - 20 meter intervals are adequate but must be adjusted to the size of the area being mapped.)
4. Make the origin the lowest mark on the baseline. Using the level sight or water level, determine the relative altitude of each marker. Record the relative altitudes along the baseline to the nearest whole meter.

Field measurement

1. Beginning at the central marker or the assigned point on the baseline, have each group sight its cross-section line along the assigned heading of the compass. The heading is assigned arbitrarily on the central-point orientation and is perpendicular to the baseline in that orientation. Students should sight and mark the line with flagging the total distance they plan to measure.
2. Beginning at the central marker or the assigned point on the baseline, direct students to use the level sight and tape measure to measure the horizontal and vertical intervals along the marked cross-section line. Students may use the record sheet from *Hill-Sides* to record their data. On the central-point orientation, the central point is the origin and has a relative altitude of zero. On the baseline orientation, the relative altitude of the starting point of the line was measured in step 4 of *Setting up the area*.

Making the map

1. On a piece of paper, have students make a scale drawing of the cross-section lines. If you use the central-point orientation, the lines will radiate from one point at the same angles as the compass headings used. If the baseline orientation were used, the cross-section lines are just lines drawn perpendicular to the baseline at the scaled interval.
2. Have students transfer the measurements from each group's record sheets. The horizontal interval is scaled and the distance measured along the corresponding line on the drawing. A mark is made on the point on the line, and the relative altitude is recorded next to it.
3. When all the measurements are recorded, have students connect the points of equal value with contour lines.
4. A final map can be made by using vellum and tracing over the contour lines with black ink. Have the students finish the maps including scale, contour interval, and compass heading.
5. The final map on vellum can be used to reproduce blue line copies to be distributed to the class.

Discussion

1. How accurately does the final map resemble the actual area?
2. Identify the different land forms on the map.
3. Do the land forms seem to have the right orientation to each other?
4. What steps in mapping could cause inaccuracies?
5. How could these inaccuracies be minimized?

Extension

Use the topographical map to make a model of the mountain surveyed. A procedure similar to the one used in T*opping Off Mount St. Helens* can be followed. Use the model to check accuracy. Check if the model appears to have the same "lay of the land."

The AIMS Program

AIMS is the acronym for "**A**ctivities **I**ntegrating **M**athematics and **S**cience." Such integration enriches learning and makes it meaningful and holistic. AIMS began as a project of Fresno Pacific College to integrate the study of mathematics and science in grades K-9, but has since expanded to include language arts, social studies, and other disciplines.

AIMS is a continuing program of the non-profit AIMS Education Foundation. It had its inception in a National Science Foundation funded program whose purpose was to explore the effectiveness of integrating mathematics and science. The project directors in cooperation with eighty elementary classroom teachers devoted two years to a thorough field-testing of the results and implications of integration.

The approach met with such positive results that the decision was made to launch a program to create instructional materials incorporating this concept. Despite the fact that thoughtful educators have long recommended an integrative approach, very little appropriate material was available in 1981 when the project began. A series of writing projects have ensued and today the AIMS Education Foundation is committed to continue the creation of new integrated activities on a permanent basis.

The AIMS program is funded through the sale of this developing series of books and proceeds from the Foundation's endowment. All net income from book and poster sales flow into a trust fund administered by the AIMS Education Foundation. Use of these funds is restricted to support of research, development, publication of new materials, and partial scholarships for classroom teachers participating in writing and field testing teams. Writers donate all their rights to the Foundation to support its on-going program. No royalties are paid to the writers.

The rationale for integration lies in the fact that science, mathematics, language arts, social studies, etc., are integrally interwoven in the real world from which it follows that they should be similarly treated in the classroom where we are preparing students to live in that world. Teachers who use the AIMS program give enthusiastic endorsement to the effectiveness of this approach.

Science encompasses the art of questioning, investigating, hypothesizing, discovering and communicating. Mathematics is the language that provides clarity, objectivity, and understanding. The language arts provide us powerful tools of communication. Many of the major contemporary societal issues stem from advancements in science and must be studied in the context of the social sciences. Therefore, it is timely that all of us take seriously a more holistic mode of educating our students. This goal motivates all who are associated with the AIMS Program. We invite you to join us in this effort.

Meaningful integration of knowledge is a major recommendation coming from the nation's professional science and mathematics associations. The American Association for the Advancement of Science in *Science for All Americans* strongly recommends the integration of mathematics, science and technology. The National Council of Teachers of Mathematics places strong emphasis on applications of mathematics such as are found in science investigations. AIMS is fully aligned with these recommendations.

Extensive field testing of AIMS investigations confirms these beneficial results.

1. Mathematics becomes more meaningful, hence more useful, when it is applied to situations that interest students.
2. The extent to which science is studied and understood is increased, with a significant economy of time, when mathematics and science are integrated.
3. There is improved quality of learning and retention, supporting the thesis that learning which is meaningful and relevant is more effective.
4. Motivation and involvement are increased dramatically as students investigate real world situations and participate actively in the process.

We invite you to become part of this classroom teacher movement by using an integrated approach to learning and sharing any suggestions you may have. The AIMS Program welcomes you!

AIMS Education Foundation Programs

A Day With AIMS

Intensive one-day workshops are offered to introduce educators to the philosophy and rationale of AIMS. Participants will discuss the methodology of AIMS and the strategies by which AIMS principles may be incorporated into curriculum. Each participant will take part in a variety of hands-on AIMS investigations to gain an understanding of such aspects as the scientific/mathematical content, classroom management, and connections with other curricular areas. The *A Day With AIMS* workshops may be offered anywhere in the United States. Necessary supplies and take-home materials are usually included in the enrollment fee.

AIMS One-Week Workshops

Throughout the nation, AIMS offers many one-week workshops each year, usually in the summer. Each workshop lasts five days and includes at least 30 hours of AIMS hands-on instruction. Participants are grouped according to the grade level(s) in which they are interested. Instructors are members of the AIMS Instructional Leadership Network. Supplies for the activities and a generous supply of take-home materials are included in the enrollment fee. Sites are selected on the basis of applications submitted by educational organizations. If chosen to host a workshop, the host agency agrees to provide specified facilities and cooperate in the promotion of the workshop. The AIMS Education Foundation supplies workshop materials as well as the travel, housing, and meals for instructors.

AIMS One-Week Fresno Pacific College Workshops

Each summer, Fresno Pacific College offers AIMS one-week workshops on the campus of Fresno Pacific College in Fresno, California. AIMS Program Directors and highly qualified members of the AIMS National Leadership Network serve as instructors.

The Science Festival and the Festival of Mathematics

Each summer, Fresno Pacific College offers a Science Festival and a Festival of Mathematics. These two-week festivals have gained national recognition as inspiring and challenging experiences, giving unique opportunities to experience hands-on mathematics and science in topical and grade level groups. Guest faculty includes some of the nation's most highly regarded mathematics and science educators. Supplies and take-home materials are included in the enrollment fee.

The AIMS Instructional Leadership Program

This is an AIMS staff development program seeking to prepare facilitators for leadership roles in science/math education in their home districts or regions. Upon successful completion of the program, trained facilitators become members of the AIMS Instructional Leadership Network, qualified to conduct AIMS workshops, teach AIMS in-service courses for college credit, and serve as AIMS consultants. Intensive training is provided in mathematics, science, processing skills, workshop management, and other relevant topics.

College Credit and Grants

Those who participate in workshops may often qualify for college credit. If the workshop takes place on the campus of Fresno Pacific College, that institution may grant appropriate credit. If the workshop takes place off-campus, arrangements can sometimes be made for credit to be granted by another college or university. In addition, the applicant's home school district is often willing to grant in-service or professional development credit. Many educators who participate in AIMS workshops are recipients of various types of educational grants, either local or national. Nationally known foundations and funding agencies have long recognized the value of AIMS mathematics and science workshops to educators. The AIMS Education Foundation encourages educators interested in attending or hosting workshops to explore the possibilities suggested above. Although the Foundation strongly supports such interest, it reminds applicants that they have the primary responsibility for fulfilling *current* requirements.

For current information regarding the programs described above, please complete the following:

Information Request

Please send current information on the items checked:

____ *Basic Information Packet* on AIMS materials
____ *Festival of Mathematics*
____ *Science Festival*
____ *AIMS Instructional Leadership Program*
____ *AIMS One-Week Fresno Pacific College Workshops*
____ *AIMS One-Week Workshops*
____ Hosting information for *A Day With AIMS* workshops
____ Hosting information for *A Week With AIMS* workshops

Name ____

Address ____
Street City State Zip

AIMS Program Publications

GRADES K-4 SERIES

Bats Incredible
Brinca de Alegria Hacia la Primavera con las Matemáticas y Ciencias
Cáete de Gusto Hacia el Otoño con la Matemáticas y Ciencias
Fall Into Math and Science
Glide Into Winter With Math and Science
Hardhatting in a Geo-World
Jawbreakers and Heart Thumpers
Overhead and Underfoot
Patine al Invierno con Matemáticas y Ciencias
Popping With Power
Primariamente Física (Revised Edition)
Primariamente Plantas
Primarily Physics (Revised Edition)
Primarily Plants
Sense-able Science
Spring Into Math and Science

GRADES K-6 SERIES

Budding Botanist
Critters
Mostly Magnets
Principalmente Imanes
Ositos Nada Más
Primarily Bears
Water Precious Water

GRADES 5-9 SERIES

Down to Earth
Electrical Connections
Conexiones Eléctricas
Finding Your Bearings (Revised Edition)
Floaters and Sinkers
From Head to Toe
Fun With Foods
Historical Connections in Mathematics, Volume I
Historical Connections in Mathematics, Volume II
Machine Shop
Math + Science, A Solution
Our Wonderful World
Out of This World (Revised Edition)
Pieces and Patterns, A Patchwork in Math and Science
Piezas y Diseños, un Mosaic de Matemáticas y Ciencias
Soap Films and Bubbles
The Sky's the Limit (Revised Edition)
Through the Eyes of the Explorers: Minds-on Math & Mapping

AIMS Duplication Rights Program

AIMS has received many requests from school districts for the purchase of unlimited duplication rights to AIMS materials. In response, the AIMS Education Foundation has formulated the program outlined below. There is a built-in flexibility which, we trust, will provide for those who use AIMS materials extensively to purchase such rights for either individual activities or entire books.

It is the goal of the AIMS Education Foundation to make its materials and programs available at reasonable cost. All income from sale of publications and duplication rights is used to support AIMS programs. Hence, strict adherence to regulations governing duplication is essential. Duplication of AIMS materials beyond limits set by copyright laws and those specified below is strictly forbidden.

Limited Duplication Rights

Any purchaser of an AIMS book may make up to *200 copies* of any activity in that book for use at *one school site*. Beyond that, rights must be purchased according to the appropriate category.

Unlimited Duplication Rights for Single Activities

An individual or school may purchase the right to make an unlimited number of copies of a single activity. The royalty is $5.00 per activity per school site.

Examples: 3 activities x 1 site x $5.00 = $15.00
9 activities x 3 sites x $5.00 = $135.00

Unlimited Duplication Rights for Whole Books

A school or district may purchase the right to make an unlimited number of copies of a single, *specified* book. The royalty is $20.00 per book per school site. This is in addition to the cost of the book.

Examples: 5 books x 1 site x $20.00 = $100.00
12 books x 10 sites x $20.00 = $2400.00

Magazine/Newsletter Duplication Rights

Members of the AIMS Education Foundation who Purchase the *AIMS* magazine/*Newsletter* are hereby granted permission to make up to 200 copies of any portion of it, provided these copies will be used for educational purposes.

Workshop Instructors' Duplication Rights

Workshop instructors may distribute to registered workshop participants a maximum of 100 copies of any article and /or 100 copies of no more than 8 activities, provided these six conditions are met:

1. Since all AIMS activities are based upon the *AIMS Model of Mathematics* and the *AIMS Model of Learning*, leaders must include in their presentations an explanation of these two models.
2. Workshop instructors must relate the AIMS activities presented to these basic explanations of the AIMS philosophy of education.
3. The copyright notice must appear on all materials distributed.
4. Instructors must provide information enabling participants to apply for membership in the AIMS Education Foundation or order books from the Foundation.
5. Instructors must inform participants of their limited duplication rights as outlined below.
6. Only student pages may be duplicated.

Written permission must be obtained for duplication beyond the limits listed above. Additional royalty payments may be required.

Workshop Participants' Rights

Those enrolled in workshops in which AIMS student activity sheets are distributed may duplicate a maximum of 35 copies or enough to use the lessons one time with one class, whichever is less. Beyond that, rights must be purchased according to the appropriate category.

Application for Duplication Rights

The purchasing agency or individual must clearly specify the following:

1. Name, address, and telephone number
2. Titles of the books for Unlimited Duplication Rights contracts
3. Titles of activities for Unlimited Duplication Rights contracts
4. Names and addresses of school sites for which duplication rights are being purchased

NOTE: Books to be duplicated must be purchased separately and are not included in the contract for Unlimited Duplication Rights.

The requested duplication rights are automatically authorized when proper payment is received, although a *Certificate of Duplication Rights* will be issued when the application is processed.

Address all correspondence to

Contract Division
AIMS Education Foundation
P.O. Box 8120
Fresno, CA 93747-8120